W9-BTF-471

THE DEMON UNDER THE MICROSCOPE

THE DEMON UNDER THE MICROSCOPE

From Battlefield Hospitals to Nazi Labs, One Doctor's Heroic Search for the World's First Miracle Drug

THOMAS HAGER

HARMONY BOOKS
NEW YORK

Published in the United States by Harmony Books, an imprint of the Crown Publishing Group, a division of Random House, Inc., New York.
www.crownpublishing.com

HARMONY BOOKS is a registered trademark and the Harmony Books colophon is a trademark of Random House, Inc.

This title may be purchased for business or promotional use or for special sales. For information, please write to: Special Markets Department, Random House, Inc., 1745 Broadway, MD 6–3, New York, NY 10019 or specialmarkets@randomhouse.com

Library of Congress Cataloging-in-Publication Data
Hager, Thomas.
The demon under the microscope : from battlefield hospitals to Nazi labs, one doctor's heroic search for the world's first miracle drug / Thomas Hager.—1st ed.
p. ; cm.
Includes bibliographical references and index.
1. Domagk, Gerhard, 1895–1964. 2. Sulphur drugs—History. 3. Antibacterial agents—History. 4. Bacterial diseases—Chemotherapy. 5. Medical scientists—Germany—Biography.
[DNLM: 1. Domagk, Gerhard, 1895–1964. 2. Physicians—Germany—Biography. 3. History, 20th Century—Germany. 4. Sulfonamides—history—Germany. WZ 100 D665h 2006] I. Title.
RM666. S9H34 2006
615'.2723—dc22 2006004510

ISBN-13: 978-1-4000-8213-1
ISBN-10:1-4000-8213-7

Printed in the United States of America

Design by Lauren Dong

10 9 8 7 6 5 4 3 2 1

First Edition

CONTENTS

ACKNOWLEDGMENTS

SPECIAL THANKS GO TO Corporate Historian Ruediger Borstel, M.A., whose help at the Bayer Archive in Leverkusen was vital to the success of this work. Michael Frings at Bayer searched out and made available scores of valuable photographs. The Bayer Corporation itself provided travel support for my research in Leverkusen, for which I thank the company in general and Thomas Reinert in particular. At the Pasteur Archives, Stéphane Kraxner provided very able and courteous assistance. At the Wellcome Library in London, I received welcome help from Helen Wakely, archivist. I relied as well on the able reference staff at the University of Oregon's Knight Library and the Oregon Health and Sciences University. Clifford Mead, head of special collections at Oregon State University, provided insightful and intelligent advice throughout the project.

This book would not have been possible without my agent, Nat Sobel, who put me together with the editors at Harmony. Special thanks here to my editor, Julia Pastore. Maureen Sugden was an exceptional copy editor. Translators were essential for understanding works in French and German, and I was assisted capably by Geraldine Poizat-Newcomb, Yasmin Staunau, Matthias Vogel, and Gerhard Spitteler, as well as my German-savvy son, Jackson Hager. Other scholars who provided support and advice included Suzanne White Junod; John H. Mather, M.D.; Ute Deichmann; Kees Gispen; Brian J. Ford; Frank Ryan; Mary Jo Nye; Charles C. Mann; and Mark L. Plummer.

As every author knows, writing a book can make you temporarily crazy. Thanks to my family—especially my patient and loving partner in life and writing, Lauren Kessler—for putting up with this one.

Tom Hager
Eugene, Oregon

As a surgeon Asklepios became so skilled in his profession that he not only saved lives but even revived the dead; for he had received from Athena the blood that had coursed through the Gorgon's veins, the left-side portion of which he used to destroy people, but that on the right he used for their salvation.

—THE LIBRARY OF APOLLODORUS

THE DEMON UNDER THE MICROSCOPE

INTRODUCTION

IN 1931, humans could fly across oceans and communicate instantaneously around the world. They studied quantum physics and practiced psychoanalysis, suffered mass advertising, got stuck in traffic jams, talked on the phone, erected skyscrapers, and worried about their weight. In Western nations people were cynical and ironic, greedy and thrill-happy, in love with movies and jazz, and enamored of all things new; they were, in most senses, thoroughly modern. But in at least one important way, they had advanced little more than prehistoric humans: They were almost helpless in the face of bacterial infection.

For thousands of years, humans had sought medicines with which they could defeat contagion, and they had slowly, painstakingly, won a few battles: some vaccines to ward off disease, a handful of antitoxins. A drug or two was available that could stop parasitic diseases once they hit, tropical maladies like malaria and sleeping sickness. But the great killers of Europe, North America, and most of Asia—pneumonia, plague, tuberculosis, diphtheria, cholera, meningitis—were caused not by parasites but by bacteria, much smaller, far different microorganisms. Nothing on earth could stop a bacterial infection once it started.

It was not for lack of trying. Like the two snakes entwined on the staff of Hermes—the symbol of Western physicians—the history of medicine comprises two twined skeins of inquiry: understanding how the body works and using that understanding to prevent its destruction. Great strides had been made in the first area. By 1931, physicians had a sophisticated knowledge of how the organs of the body and the systems they created—digestive, hormonal, nervous, and so forth—cooperated to create health. They had made a good start on moving from the level of organs and tissues down into the intricacies of molecular biology (a term invented in the mid-1930s). They knew a great deal about what happened to those organs, tissues, and systems when they were hit by disease. But there the knowledge stopped. The great prize eluded them.

That prize, sought since antiquity, was called panacea, a mythical substance that could cure the sick and raise the dead. (*Panacea,* literally "all-healing," was also the name of the ancient Greek goddess of health, daughter of the physician-god Asklepios.) The Egyptians hoped that the art of mummification would lead them to it. The Greeks sang of it. Medieval monks believed that it could be reached through holy relics. Alchemists sought it as the Philosopher's Stone, which would not only turn base metals into gold and transform the soul of the seeker but would cure all maladies. It appeared in legends and fairy tales as Achilles' spear, Aladdin's ring, Fierabras's balsam, Medea's kettle, and Prince Ahmed's apple. Before the scientists took over, generations of magicians, mages, scholars, and snake-oil salesmen had pursued panacea. But no one had found it. Once a bacterial disease took hold in the body, humans in 1931 were as much the prey of the invisible killers as they had been since the beginning of history.

All that was about to change.

I STUMBLED across this story—which I now consider one of the most important in modern history—quite by accident, a method appropriate for a discovery comprising equal parts skill, mistake, luck, and bullheaded idealism. A reformed scientist—I studied medical microbiology for years before deciding that I would rather write about the beautiful, painstaking rigor of bench science than actually do it—I was happily thumbing through my dog-eared, spine-sprung copy of *Asimov's Biographical Encyclopedia of Science and Technology: The Lives and Achievements of 1510 Great Scientists from Ancient Times to the Present Chronologically Arranged,* a delightful candy store of a book for science buffs. I started doing what Isaac Asimov intended his readers to do, I think, when he wrote the book with an emphasis on generous cross-referencing: that is, I began linking the work of one scientist to another, tracing currents of thought across nations and through time. I found my way to the entry on Emil von Behring, the stiff-necked Prussian bacteriologist, which led me to Paul Ehrlich, the Man with the Blue Fingers, whose entry contained a reference to a scientist I had never heard of before, a German physician named Gerhard Domagk. Reading the brief entry on Domagk was the first step on a two-year journey that resulted in this book.

I was intrigued not so much by the man—although Domagk grew into a more interesting character the more I learned—as by the ways in which his discovery was embedded in and affected so much of what we take for granted in modern medicine. Ours is an age of science; this is an archetypal story of our time.

I am part of that great demographic bulge, the World War II "Baby Boom" generation, which was the first in history to benefit from birth from the discovery of antibiotics. The impact of this discovery is difficult to overstate. If my parents came down with an ear infection as babies, they were treated with bed rest, painkillers, and sympathy. If I came down with an ear infection as a baby, I got antibiotics. If a cold turned into bronchitis, my parents got more bed rest and anxious vigilance; I got antibiotics. People in my parents' generation, as children, could and all too often did die from strep throats, infected cuts, scarlet fever, meningitis, pneumonia, or any number of infectious diseases. I and my classmates survived because of antibiotics. My parents as children, and their parents before them, lost friends and relatives, often at very early ages, to bacterial epidemics that swept through American cities every fall and winter, killing tens of thousands. The suddenness and inevitability of these epidemic deaths, facts of life before the 1930s, were for me historical curiosities, artifacts of another age. Antibiotics virtually eliminated them. In many cases, much-feared diseases of my grandparents' day—erysipelas, childbed fever, cellulitis—had become so rare they were nearly extinct. I never heard the names.

Nor did I understand the word "physician" in the same way. To my grandparents a physician was a poorly paid, selfless caregiver who made house calls, kept vigil over the sick, and comforted the families. To me a physician was a wealthy technician in a white coat who did a quick exam in the office and wrote a prescription. Prescriptions had changed, too. Before 1935, narcotics were practically the only drugs that required a prescription. Everything else was sold over the counter. Today virtually every powerful medicine requires a prescription. Before 1935, patent medicines were one of the biggest businesses in the United States. Today they no longer exist. What had happened?

Sulfa happened. It started in the mid-1930s, with a series of findings made in Germany and France, discoveries that were at the time hailed as "the miracle of miracles" in modern medicine, advances that secured humans their first effective way to stop bacterial infections once they started. The work then spread to Great Britain and the United States, where tests of the still-experimental drug on humans, including the son of the president of the United States, confirmed its power. The story became stranger and more colorful the more I researched it, the characters and stories more intense, featuring Congo Red and methylene blue, the Holy Fire of Viennes, vats of Scottish tar, Roehl's infected carbuncle and Duisberg's Council of the Gods, Queen Victoria's armpit and St. Anthony's bones, impossibly small animals and impossibly huge business cartels.

This is that story.

A NOTE ON USAGE: The family of medicines now generally called sulfa drugs includes thousands of related molecules that have been called many other, often more specific things in the scientific literature. I use "sulfa" throughout this book as a generic term for any medicine whose activity can be traced back to a relatively simple set of atoms called sulfanilamide (sulfa is a relatively common nickname for sulfanilamide and its related substances). I use "sulfa" and "sulfanilamide" interchangeably; the phrase "sulfa drugs" as I use it in this book encompasses the myriad sulfanilamide-containing substances created after the discovery of its therapeutic powers, including those substances that physicians and chemists call sulfonamide drugs. The term "antibiotic" is defined two different ways in the literature; the first, more strict, definition dictates that for a medicine to be called an antibiotic it must be produced by a living microorganism, as penicillin is produced by a mold. Thus some experts refuse to call purely synthetic chemicals like the sulfa drugs, made in laboratories rather than by nature, antibiotics. The more reasonable definition, to my way of thinking and that of a number of medical experts, instead ties the word "antibiotic" to what a substance does rather than where it comes from. Used in this way, an antibiotic is any substance that can selectively destroy a spectrum of bacteria within the body without significantly damaging the body itself. This is how I employ the word throughout the book. By this definition, sulfa was the world's first antibiotic.

PROLOGUE

JOHN J. MOORHEAD, a lean and intense New York surgeon, was thrilled to be in Hawaii. One of the leading trauma surgeons in the United States and an expert in wound treatment, Moorhead had been invited by the Honolulu County Medical Society to speak to doctors and nurses in December 1941. He gratefully accepted. It provided him a welcome chance to trade the winter winds of Manhattan for the tropical breezes of Oahu. He arrived on schedule and spoke to two large groups on the subject of trauma surgery. His second talk, "Treatment of Wounds, Civil and Military," proved especially timely. America was not at war yet, but rumors were flying. Everyone was talking about the possibility of a Japanese attack on the Hawaiian Islands.

On his way to deliver a third lecture on the morning of December 7, Moorhead and his driver heard an announcer on the car radio say the U.S. naval base at Pearl Harbor had been attacked. "You hear all kinds of things out here," his driver said, and they continued to the lecture hall. There Moorhead found only about fifty doctors and nurses in their seats instead of the three hundred he had expected. Moorhead went ahead anyway; he had just started to speak when a man burst into the hall and asked all physicians in the audience to come immediately to Tripler General Hospital, Hawaii's largest military medical facility. The attack had been real. The hall emptied. Moorhead, too, reported to Tripler.

He and his driver arrived to find stretchers covering the lawn in front of Tripler's main building. A stream of ambulances, trucks, and private cars had been making the four- or five-mile trip back and forth from Pearl Harbor, dropping off the wounded wherever they could find space under the flowering trees. Aides were running from man to man, improvising tourniquets from belts, gas-mask cords, pistol holsters, and strips of sheets. Moorhead went inside, was informed that he was now a colonel in the Army Medical Corps, and scrubbed up. Eight surgical teams were quickly formed from arriving physicians and nurses, most of them civilians. They shared three operating rooms and worked without pause for the next eleven hours, passing medical instruments from room to room, performing hundreds of operations, filling garbage barrels with amputated limbs. One surgical saw was used and sterilized so many times it remained hot to the touch all day.

The worst of the early cases came from Hickham Air Field, where a Japanese bomb had hit the mess hall at breakfast time. Thirty-five young airmen had been killed outright. Dozens more arrived at Tripler with "wounds too terrible to describe," a nurse said. The bomb had blown out the mess hall's walls and windows; the men's wounds were contaminated with shrapnel, shards of glass, bits of mortar and brick, and partially digested food.

Six years earlier this would have been a recipe for disaster. It was a military fact of life then that regardless of the care of the physician and the success of the initial operation, wounds often got infected, and many times infected wounds killed the patient. The dirtier the wound, the greater the risk. Once a wound infection started, almost nothing could be done. There were no antibiotics. In America's last war, World War I, infected wounds had killed hundreds of thousands of men—more soldiers, by one estimate, than died by enemy bullets. In World War II, however, the numbers would be so low that wound infections no longer presented a major medical problem.

The difference was a new family of drugs—sulfa drugs—made public in the United States just five years before the attack on Pearl Harbor. They were the most effective, most important medicines ever discovered. They had already saved the life of President Roosevelt's son and those of tens of thousands of other patients around the world.

They were revolutionizing medicine. At Pearl Harbor they received their toughest test yet.

Luckily, Tripler Hospital was well stocked with sulfa. Every wounded man who could swallow was given sulfa tablets. The rest got shots or had it sprinkled into their wounds. If someone's abdomen had been opened, the physicians cleaned it out as well as they could, stopped the bleeding, and packed the cavity with sulfa. When the burn patients began arriving from the harbor, sailors who had been in water covered with blazing oil—"roasted men," as one observer put it—their burns were cleaned, the dead tissue stripped away, and the injured area frosted with sulfa.

"The casualties were numerous, varied, and severe," Moorhead wrote in his official report. It was a grisly test of the new medicine's power. It had been just thirty-six hours since he had delivered lectures on the best methods for wound treatment. Now his advice was being used, step by step, to handle the men at Tripler: Cleanse the wound with soap and water; carve away all dead and damaged flesh (an old World War I method called debridement) until tissue is reached that is normal color or, if muscle, capable of contraction; tie off everything bleeding; pack the wound with sulfa; leave it covered with gauze bandages, unsutured for three days or until it is certain that no infection has set in; then close if appropriate. During the entire recovery period, every patient was to be given a gram of sulfa every four hours. "No one then thought," Moorhead wrote, "that these principles of treatment were so soon to be put to a large scale test in a proving ground only a short distance from the lecture platform."

For the next ten days, Moorhead and the other physicians tracked their patients' progress. The most feared type of wound infection, gas gangrene, was a virtual death sentence for wounded soldiers. At Tripler the gas gangrene cases were isolated in their own ward, their wounds opened and cleaned a second time, then sulfa repacked into the wound. More sulfa was given by mouth. Every patient with gas gangrene recovered, without a single additional amputation. Of the wounded at Pearl Harbor who did not die directly from the trauma of the injury during the first week, not a single one was subsequently lost to infection. Nothing close to it had ever been seen in the annals of military medicine. The official U.S. history of World War II later credited the low mortality rate to "skilled surgery and the use of sulfa drugs." John Moorhead returned to New York with a deep appreciation for sulfa. His feelings were shared by every physician who treated wounds during the war.

Ironically, the medicine that helped the United States in World War II was discovered in a laboratory in Germany, in the year Hitler came to power, by a corporation whose executives would be put on trial for war crimes at Nuremberg. It would change the way new drugs are developed, approved, and sold. It would transform the way physicians deal with patients. It would usher in the era of antibiotics and lay the foundation for what we now consider modern medicine.

All of this would result from a combination of corporate strategy, individual idealism, careful planning, lucky breaks, cynicism, heroism, greed, incredibly hard work, and one central, overarching, and mistaken idea.

The story of sulfa's reign as the world's leading wonder drug is a short one. It burst onto the global stage in the mid-1930s, created enormous excitement, then all but disappeared just ten years later. Yet in that short time it changed everything.

I

The Hunt

The art of medicine consists of amusing the patient while nature cures the disease.

Voltaire

CHAPTER ONE

GERHARD DOMAGK looked at the blood soaking his tunic. It was 1914, a few days before Christmas. The German army had just finished an artillery barrage. Domagk's unit had been sent in, the young men and their officers walking slowly through the yellowing grass toward a Polish farmhouse, their breath showing white, when shots came from somewhere to their left. Domagk saw the officer nearest him fall. Then he felt a blow to his head. His helmet flew off and landed somewhere in the grass. His chest felt hot. When he looked down, he saw the blood. He had attended a single term of medical school before joining the army and knew enough to give himself a quick exam. He found no wound on his body. Then he discovered the source. Blood was streaming from his head, down his neck, and onto his shirt. He explored his scalp gently with his fingers. Hard to say how bad the gash was, but it had probably opened when the bullet knocked his helmet off. He bandaged himself with a large handkerchief. Then he passed out. When he awoke, he was jolting through trees in a farmer's cart toward what had been a church, now a German field hospital, where he was examined, his bleeding stopped, and his wound dressed. When the staff decided that it appeared likely he would survive, he was packed onto a train to Berlin, to recuperate in a central hospital. The wound did not look serious, but there was no way to know if there would be permanent brain damage. Time would tell.

The blow to his head did not change Domagk's mind about the war. He, like most of his fellow university students, had been infected and rendered mildly delirious during the epidemic of patriotic fever that swept Germany in the summer of 1914. The tall, thin boy volunteered for service with more than a dozen of his classmates and friends soon after war was declared. They were inducted as a group into the Leibgrenadier Regiment of Frankfort on the Oder, a unit specializing in the use of grenades. They were given a few weeks of cursory training. Then they were loaded onto a train for Flanders.

They were young and full of energy, eager to join Germany's march, giddy with visions of a short, glorious war. Domagk, the son of a village schoolmaster, was eighteen years old and ready for adventure. He was also a young gentleman who brought his lute to training camp and played folk tunes around the campfire. He wanted to take the instrument with him to the front. When his officers told him that regulations forbade it, he dismantled it, sent the body back to his parents, and kept the neck attached to his knapsack as a memento. Inside the knapsack he carried a photo of his village sweetheart dressed in her white Communion gown.

Now, months later, he was beginning to miss his home, Lagow in the lake country of far eastern Germany. Picturesque and quiet, Lagow became the source of ever-sunnier memories the longer he spent in the army: cannonballing into the river below the mill; a swarm of children flying out of school at the end of the day; a group of friends concocting homemade gunpowder; sneaking his first cigar; the taste of a ripe pear in late summer. He spent his nineteenth birthday in the trenches of Flanders under fire from British ships, huddled in the dirt, "the heavens lit," he wrote his parents, "from burning villages." The glory of war began to fade. He and his comrades were soaked by freezing autumn rains, exhausted, starving, their uniforms caked with muck. Once while digging for drinking water, they broke open an abscess in the earth, a cache of rotting French soldiers, men killed and buried, he figured, by his own unit's grenades.

The Germans were dug in near the Belgian coastal town of Nieuport, where in late October the Leibgrenadier Regiment of Frankfort on the Oder was ordered to participate in a massive attack. Their officers told them that following a 4:00 A.M. German artillery barrage they would charge forward from their trenches and drive the enemy from *their* trenches. The young men synchronized their watches. They wrote last letters home and put them in their pockets, promising each other that the living would deliver them for the dead. They waited for what seemed a very long time in the dark, listening to shells screaming overhead, watching the flashes.

When the barrage stopped, the young German soldiers struggled and slipped out of their holes. They slogged through a football field's length of mud before they started falling, then heard the chattering of machine guns at short range, each one firing as many bullets as 250 rifle-equipped soldiers. Most of the boys Domagk had joined with were dead within a few seconds. The rest ran. Domagk later figured that only he and two or three others out of his group of fifteen student volunteers survived the battle alive and unwounded. They learned later that their charge was part of a huge failed offensive in which the Germans lost 135,000 soldiers, many of them recent university students, in the course of four weeks of fierce fighting. The British called it the First Battle of Ypres. The Germans called it *Kindermord:* "The Massacre of the Innocents."

Too ripped up to fight any longer in Flanders, Domagk and what remained of the Leibgrenadiers were transferred to the Eastern Front. A few weeks later, he lost his helmet near the Polish farmhouse. When he began to gather his senses about him in a Berlin hospital room, he discovered that his knapsack was gone, along with the neck of his lute and the photo of his sweetheart. All he now had of his childhood were memories. He remembered his father sitting at the window, waiting for the lamp man. The gas streetlights in Lagow were lit every evening by the lamp man, who came came by with his white horse. Then one day the man stopped coming. When Gerhard's father explained to him that the lamp man had been delayed because his horse was sick, the young boy was stricken by the idea. "I said at the end of the evening prayer with my mother, 'Good God, please make the lamp man come again,'" he remembered. "'Make his horse better again.'"

The German army hospital administration in Berlin, on reviewing their records, found that wounded Leibgrenadier Gerhard Domagk had attended a bit of medical school. It was decided that rather than send him back to the front lines, they would set him to the task of providing medical care for the wounded. Domagk was placed in a training program for medical assistants, one of hundreds of novices hastily pressed into service. After a few weeks of first-aid training, he was sent back to the Eastern Front, through Kraków to a field hospital in the Ukraine. He was fascinated by his trip through "the culture of the East," as he called it, the lands of Germany's destiny, the "beautiful but dirty streets," the Jews with "caftans reaching over their long boots and their corkscrew-like curls hanging down from their temples." He was especially impressed by the architecture he saw.

Flanders had been bad, but the Eastern Front was in many ways worse, especially when it came to medical care. The German casualties were just as heavy, but the hospitals were cruder, doctors fewer, supplies scarcer. The field hospital to which Domagk was assigned was stark, a farm in the middle of the woods roughly converted into a care facility with tents for wards and a barn for an operating room. Every day a miscellany of ambulances, cars, trucks, and farm carts arrived, disgorged their loads of quiet, white-faced wounded, and left for more. There was a constant, deep rumble from big guns a few miles away.

They were seeing wounds no one had ever seen before, thanks to the advance of military and industrial science. Newly deployed and unprecedentedly powerful weapons—artillery that could shoot shells 120 kilometers, high-explosive shells like the giant "Jack Johnsons" that geysered black earth a hundred feet in the air, airplanes and aerial bombs, tanks and poison gas—were slaughtering men at a rate and in ways unimaginable a few years earlier. In previous wars men had been shot or stabbed. Now they were blown to bits. The new weapons changed both the manner of fighting—more trenches, fewer cavalry charges—and what happened after. Because of the new weapons, the number of dead and wounded on both sides was staggering. During the entire Franco-Prussian War in the 1870s, a total of a quarter of a million men were killed and wounded on both sides over ten months of battle—roughly the same total number of killed and wounded at the First Battle of Ypres alone. Military leaders realized within a few months of the war's start that they needed to quickly expand their medical services. Anyone with any medical ability was pressed into service in the rapidly growing network of hospitals. That was how Domagk ended up in the woods of the Ukraine.

One of his jobs was to sort the new arrivals, separating the obvious infectious cases, cholera especially, and moving them out of camp as quickly as possible; setting aside those whose wounds were least severe, those with the best chance of recovery, and sending them immediately into the queue for surgery; then arranging for the remainder, the severely wounded, those with parts of their heads blown away, the cases with burst intestines, those no longer responsive. The worst were put under the straw. Comforted as much as possible, of course, given water, told their turn would come, to keep up their spirits, that the straw would keep them warm. They were generally in shock anyway, pulses thready, color graying, skin growing cold. They would last a night or two at most. Another of his jobs was to go out in the morning, pull the straw away, and make a final sorting. The living were reexamined. The dead went into that day's mass grave.

With the grisly morning chore out of the way, aides like Domagk were often given strip-and-clean duty. The wounded arrived filthy, their shredded uniforms crusted with dried blood and dirt. Some had been wounded for days, lying in water pooled in the bottom of bomb craters. Many were infested with lice. The young aides cut away the uniforms, peeled them off and burned them, then tried to wash the men as well as they could, working around the open wounds and broken bones. Most of the men were surprisingly silent and stoic, showing their strength. Others moaned or sometimes screamed.

That done, Domagk was often called to help in the operating room. The surgeons worked in shifts twenty-four hours a day, the operating theaters in the barn, lit with hissing carbide lamps at night, or in tents. Everything was in short supply—morphine, anesthetic, sterilization equipment, surgical instruments. It was impossible to keep the area clean. Gowning and masking were rare; assistants like Domagk worked bare-handed in their uniforms. "We could hardly keep the flies away," he remembered. All the medical staff were dizzy from exhaustion and from breathing the ether and chloroform they dripped onto pads and held over the mouths and noses of the wounded. The physicians and aides fled the operating areas whenever they could, napping outside even in the rain to get away from the insects and the fumes. Get what sleep you can; sort the new arrivals; strip and clean; pull the bodies from the straw; assist in surgery; fall into bed; get what sleep you can; sort the new arrivals. . . . This was Domagk's life for two years.

He was smart, young, and strong, he followed orders and worked hard. He earned a reputation as a reliable, steady man. And he was unusually observant. Some people had a feeling for medicine, a caring personality, or—like the surgeons—a combination of hand-eye coordination and steady nerves. Domagk had the ability to *see*. He watched everything, noted slight variations, quietly filed it all away. He witnessed more operations in two years than many surgeons see in a lifetime; helped set compound fractures, the bones bristling through the skin; used magnets to search for pieces of shrapnel; watched the surgeons run their fingers down the insides of intestines, probing for holes; assisted with countless amputations; threw the severed arms and legs onto a growing pile in a side room. The physicians came to trust him as he spent more and more time in the operating room.

Domagk for his part came to have a profound respect for what could be learned by cutting through the skin. He saw what lay beneath, saw what could be fixed with knives and thread and what could not. He saw that surgery could sometimes work miracles, and he learned about other things in medicine that did not work. Infectious disease was the bane of every military facility. Cholera was an especially severe problem in close-packed military camps. Around a campfire one evening, an old officer serving in Domagk's hospital told the young assistants about an epidemic of cholera that had raced through Hamburg fifteen years earlier, killing thousands in the city; a terrible disease, he said, the man still unsettled by the scope and swiftness of the contagion he had witnessed. He then told the young aides a bit of folk wisdom: Alcohol, he said, generously imbibed, could ward off cholera. The only thing they had in camp, apart from the pure alcohol sequestered to clean wounds, was red wine, so all the young men started drinking as much as they could get. Then, a few evenings after he was told the story, Domagk remembered, the old officer started complaining of a pain in his calves. A few hours after that, the older man was dead—from cholera. Folk wisdom, Domagk learned, was little protection against disease.

The most important thing that Domagk observed in the Ukraine was that even the most heroic and seemingly successful surgeries could go completely wrong a few days later. A soldier could wake one morning to find his carefully closed incisions, which had been fine the day before, now swollen, red, and painful. The edges, perhaps, had started to split open. Sometimes a foul-smelling, dark liquid oozed out. The skin around the wound began to take on a "a curious half-jellified, half-mummified look," as one physician described it. These were signs of what military physicians feared most in their postoperative cases: *Gasbrand,* the Germans called it. Gas gangrene. The doctors knew what caused *Gasbrand*—an infection by bacteria—and they knew how it progressed, the microscopic invaders eating away and essentially rotting the muscle tissue, releasing as they went both toxins that poisoned the patient and a gas that ballooned his wounds. The condition could be confirmed by running a finger along the soldier's skin: If the skin crackled from the tiny bubbles of gas under the surface, it was gas gangrene. There was nothing much that could be done. In severe cases the only thing left to physicians was to keep cutting. Second, third, even fourth amputations were attempted as surgeons tried to remove tissue well ahead of the advancing bacteria. The soldiers who survived the pain and shock of the repeated surgeries emerged horribly disfigured. But not many survived in the early days of the war, especially on the Eastern Front. Once gas gangrene was under way, the bacteria almost always won. Some patients fought it, railing and ranting for a day or two. Then they usually gave up, went silent and pale, temperature dropping, lips bluish. A day or two later, they quietly died of "green-black gangrene," one historian wrote, "which emptied surgical wards into the graveyard."

Gas gangrene was furiously contagious. Once loose in a postoperative ward, it could kill half the wounded patients in a few weeks. So, once diagnosed, gangrene patients were immediately isolated in their own wards, silent places where the air seemed close and greenish, packed with doomed men. Going into a *Gasbrand* ward was like walking into a swamp. The first reaction of new visitors was to hold their breath. The smell of rot was overpowering; those nurses and attendants whose duty it was to work there found that after a few days they could no longer wash the smell out of their skin or clothes. Gas gangrene was the deadliest but not the only type of wound infection. Other bacteria could infect a wound, alone or in concert, causing variations on the progress of the disease, any or all of which could kill injured soldiers. Taken together, wound infections accounted for an army of German dead—between one hundred thousand and two hundred thousand soldier/victims—during World War I. "It was a wonder to me," Domagk remembered of his days in the field hospital, "that it was possible at all to conduct any large operation under these circumstances successfully and sufficiently free of germs that the patient didn't die from a wound infection."

Over the course of his two years in the Ukraine, Domagk grew numb to the exhausting, repetitive horror. He began to look forward to any little variation: a request to mail a letter in a nearby town, a cigarette butt, a ride on a bicycle, a walk in the woods. And he slowly changed. By the time the war ended, the enthusiastic student had become a pensive, reserved young man, manner controlled, lips thin, smiles rare, blond hair cut short above a high intellectual's forehead, eyes blue and deep-set, with a look that after the war his friends and colleagues often described as sad or haunted.

"The total seriousness of life and death gripped us," he remembered. "It was only now that I knew why being a doctor was such a beautiful thing—and also what it could require of you." The wounds themselves he accepted as the results of war. But the infections that followed—surely science could do something to stop those. He focused on the bacteria, his personal demons, "these terrible enemies of man that murder him maliciously and treacherously without giving him a chance."

"I swore before God and myself," he later wrote, "to counter this destructive madness."

CHAPTER TWO

A FEW hundred miles away in France, just before Domagk's unit was decimated at Ypres, another, far more powerful man had also dedicated himself to beating wound infections.

Sir Almroth Wright took one sniff of the badly lit lower-level rooms he had been given in the Boulogne Casino, called for Sergeant William Clayden—Clayden, everything Sir Almroth was not, British army through and through, always regulation, spit-and-polish—and instructed him to pour cresol down the drains several times a day. Perhaps that would stop the intolerable reek. It came, Sir Almroth figured, from the decomposition of draining blood. It was the smell of rot, very similar to the horrible smell of gas gangrene. No one could work under these circumstances. Certainly not Sir Almroth Wright, veteran warrior of science and infamous curmudgeon. Already in his fifties when called back into military service, a bearlike man with a graying walrus mustache and waning respect for anything having to do with the military, Sir Almroth was not about to change his nature. He spoke his mind on every subject, from the absurdity of women's suffrage to the insanity of wartime medicine. And he spoke it publicly, to his colleagues, to govermental agencies, and through letters to the *Times,* where he had already complained early in the war that "an army, on going out into active service, goes from the sanitary conditions of civilization straight back to those of barbarism." That was what he was seeing now in British General Hospital No. 13 in Boulogne, France, showcase of his nation's military hospital system in support of the British Expeditionary Force. The BEF was expected to win quick victories and come home. The medical support system was set up to heal the warriors and had been designed rationally for maximum effectiveness: field stations near the front lines to quickly patch the wounded and rush them back to behind-the-lines hospitals like Boulogne, where they would get the best care possible before being shipped home to England for final recuperation. Boulogne was intended to be a modern repair shop for shattered men. Instead Sir Almroth when he arrived found a throwback to the septic conditions of the Middle Ages. The place smelled like it anyway, no matter how much cresol Sergeant Clayden poured down the drains.

The British had created the hospital by converting what had been the town's casino, swathing the ornate chandeliers in white linen, replacing the gaming tables with phalanxes of beds, and transforming the private card rooms—the *salles privés*—into operating rooms. The bedding was boiled and bleached, the wards scrubbed spotless, the nurses were plentiful, the physicians included some of the empire's best. And the patients were dying in droves.

The problem here, as it was in the Ukraine, was the rampant infection of wounds. British soldiers—regardless that they received the best care in the world—were dying almost as fast as the German soldiers were in Domagk's filthy field station. As the flood of wounded poured in, it soon became clear that more British soldiers were dying from wound infections than from enemy bullets. Something had to be done. So the army turned to Sir Almroth Wright, hoping that England's greatest bacteriologist and expert in infectious disease could solve the problem.

It was something of a gamble. The military administrators knew that there was no love lost between Sir Almroth and the army. The scientist had already resigned his commission once, a decade earlier, very publicly, creating what to the military administration had seemed like a totally unnecessary scandal in the press. The problem then had been hesitation. The British generals had balked at inoculating troops leaving for the Boer War with a vaccine Wright had created and which he swore would provide protection against typhoid fever. The generals were worried that Wright's vaccine was unproven, that British soldiers would be used as guinea pigs, that side effects would put more soldiers out of action than the medicine would save. They were correct on the first two points at least. They sent the men off without vaccinations, and fifteen thousand of them died of typhoid in South Africa, where disease killed twice as many British soldiers as the Boers did. After that, Wright was allowed to test his vaccine on British soldiers in India, where it proved a success. It was a public embarrassment. Wright was not shy about pointing out that those thousands of soldier typhoid victims might have lived but for the bullheadedness of their commanders, and he resigned his commission in 1902. His successful vaccine, however, earned him a knighthood and his vindication: During the First World War, millions of doses of his antityphoid vaccine were given to British troops with spectacular results. Typhoid was negligible in Flanders compared to what they had seen in the Boer War.

That success seemed to give Sir Almroth a maddening sense that he was always right about everything. After leaving the military, he had gone off and started a vaccine research laboratory in London, at St. Mary's Hospital—the Inoculation Department, he called it—and did his best to forget about the army. Sir Almroth was not the military type in any case. He was a lover of poetry, a freethinking Irishman who took tea with George Bernard Shaw (Sir Almroth, it was said, was the model for the character of Sir Colenso Ridgeon in Shaw's *The Doctor's Dilemma*), a man who demanded that everything be proved to work scientifically and to his personal satisfaction before he would implement it. In other words, he was unlikely to take orders. There was no guarantee that even in a time of war he would return to service. But the potential renown associated with solving the wound-infection problem far outweighed the risks of putting Sir Almroth in a uniform. The approach was made by Sir Alfred Keogh himself, director-general of Army Medical Services, who used as bait the combination of a commission for Wright as a colonel, an appeal to his humanitarian impulses, and the promise of substantial research funding. As it turned out, he did not need to try very hard. Sir Almroth, among his many other characteristics, was a patriot. He readily agreed for the good of his nation to set up a laboratory for wound-infection research in Boulogne.

True to form, however, once he got to Boulogne, the Old Man, as his research assistants affectionately called him, refused to shine his buttons or tuck in his shirt. He was negligent when it came to saluting. He was the despair of Sergeant Clayden. The low point came when Sir Almroth walked into the hospital one day with a large tear in the seat of his pants. Clayden, horrified, hid Sir Almroth in a side room while he made repairs. Wright was simply not regular army.

But he was brilliant. It was likely that Sir Almroth knew more about the causes and cures of infections than anyone in Britain. Now he intended to use this opportunity—the chance to create the world's first dedicated laboratory for military medicine—to quickly solve the problem of wound infections. He thought at first that it could be done relatively easily by using the same techniques he had used to solve the typhoid problem. All he had to do was find out which bacteria were causing wound infections, create a vaccine against them, and inoculate the soldiers. Problem solved.

Once on the ground in Boulogne, he surveyed the small amount of space he had been given in the lower floors, sniffed the air, and started demanding changes. He refused to stay in the stink and talked himself into space on the top floor, in a former fencing school. This, too, was less than perfect. There was no electricity up there, or proper gas, or heat, or even running water, but it was large and airy, and there was good natural light. Here he and his crack team of researchers—young men he imported from his lab in London, saving them from becoming cannon fodder on the front lines—began putting together a research facility. The place they had come from, the Inoculation Department at St. Mary's, had been so productive, so successful in advancing basic knowledge and making vaccines that they had nicknamed themselves "the House of Lords." His assistants included some of the nation's brightest young scientists, such as Alexander Fleming, who would later discover the medical properties of penicillin, and Leonard Colebrook, who would become one of England's most renowned physicians.

In Boulogne they set to work turning the fencing school into a laboratory. They divided the central arena with rough wooden walls, installed oil heaters for the winter, created lab benches out of spare tables, scavenged chairs and glassware—Sergeant Clayden proving himself a talented scrounger—substituted spirit lamps for gas flames to heat their experiments, and installed a makeshift water system with a tank on the roof. Soon every surface was packed with microscopes and racks of test tubes. Sir Almroth was famous for inventing lab techniques on the fly (he perfected the art of sterilizing needles in hot olive oil, for instance, which he found worked admirably if it was brought just to the point where it browned bread crumbs). If they needed specialized equipment, they made it out of whatever was available—fashioning a glassblowing apparatus out of an old gas can and a foot bellows, for instance. Within a few weeks, in the late fall of 1914, they were doing first-rate science.

The only problem was that none of the vaccines they tested were able to prevent or cure wound infections. Instead of a quick and brilliant victory, after a few months at Boulogne, Sir Almroth found himself in the equivalent of a trench war. "We have just got to begin at the beginning instead of trying short-cuts to cures," Wright said. He took several steps back, refocusing his laboratory on a systematic study of exactly what happened during wound infections, hoping that if they could discover precisely where the infecting bacteria came from, how infections started and how they proceeded, and how the body fought back, a point could be found to attack and defeat the infection.

Their first discovery was that the British had prepared for the wrong war. The empire's approach to battlefield medicine was based on what had been learned during the Boer War in South Africa fifteen years earlier. There the British physicians had carefully treated wounds with the most modern and systematically applied antiseptics, washes and ointments, topical medicines that killed bacteria on contact, and sterile surgical techniques. It seemed to work. Wound infections had not presented a major military problem in the Boer War. The doctors thought they had solved the problem.

They were wrong. Their techniques worked only because of the nature of the South African dirt and the types of wounds the men suffered. The Boer War took place mostly in the dry, rocky veld; British soldiers were generally injured by high-velocity Mauser bullets shot at a distance by Boer farmers. The wounds were relatively simple and clean. Treatment with a bit of antiseptic and the right dressing led to quick cures. Flash forward more than a decade to Flanders, to European farmland soaked with fall rains, and to wounded arriving shredded by shrapnel, and none of the African techniques worked. The explosive shells of World War I drove muddy, richly manured soil deep into ragged cavities in the flesh. Like the German soldiers in the Ukraine, the British wounded in Flanders often fell into stagnant water pooled in the bottoms of bomb craters, and there they would lie for hours, sometimes days, before evacuation. In Flanders there was no such thing as a clean wound. Wright's group found that the dirt in the wounds was alive with fecal bacteria common to horses and cows. They found that it was impossible to clean the wounds effectively. They found most wounds heavily colonized within a day or two by a dozen kinds of barnyard bacteria, several of which were able to cause deadly infections, including the germs that caused tetanus and gas gangrene; various strains of *Staphylococcus* and *Streptococcus;* a miscellany of other fecal and skin bacteria. By the time a doctor saw the soldiers, it was already too late as far as infection went; the wounds were dangerously septic from the very start. This, Wright believed, was why his vaccines did not work: There were too many kinds of bacteria involved, there too many targets to attack even with a mixed vaccine, the bacteria were too well established by the time the vaccine could be administered.

Then Wright's group showed that gangrene infections were the result of a stepwise process. Through a series of what one historian calls "very elegant experiments" using a seemingly limitless number of wounded men as, in a sense, laboratory animals, they tracked the progress of infection, gathering daily samples of tissue and fluids, assaying the types and numbers of bacteria, measuring the body's immune response by counting the numbers of white blood cells present at various sites in the wound, then matching their findings to the patients' general health. They found tetanus bacteria growing in the wounds of a third of the patients, strep in half of them, and the organism that causes gas gangrene in 90 percent. They found that the infections often started with the men's soiled clothing: The shells blasted muddy cloth into the wound, giving the bacteria in the dirt an excellent place to start growing. They drew blood and mixed it with feces, simulating the bacterial array found in the fields, then put the mixture in a special flask they made by heating the bottom of a test tube in a flame until it was soft, and drawing the glass out to a half dozen points, making hiding spots for bacteria, a simulacrum of a deep, ragged wound. When they added antiseptics to the flask, they found that even enormous doses failed to kill bacteria in those hiding spots. Dousing deep wounds with even the strongest antiseptics, in other words, was not going to stop wound infections in Flanders. They found that a few bacteria always escaped in deep crevices, and that the human body quickly washed the antiseptics away in any case, "quenching" their power to kill bacteria with blood serum and lymph. Because strong antiseptics killed human cells just as well as they killed bacteria, applying them to deep wounds was actually counterproductive, because the antiseptics killed many of the body's key defenders, white blood cells called leucocytes, along with the bacteria. It was essential to keep those white blood cells alive and active. "The leucocyte is the best antiseptic," Sir Almroth would tell any surgeon who would listen to him. His tests showed that the antiseptics they were pouring into wounds were not only missing a lot of bacteria but were making things worse by throwing off the proper immune response. In order to kill every dangerous germ in a deep wound, Sir Almroth figured, it would be necessary to pump in enough antiseptic to kill the patient.

So his group focused on ways to keep the immune system as healthy as possible. They invented new techniques as they went, instruments for probing and sampling wounds (including the unfortunately named "lymph leech") and for testing for and analyzing bacteria and white blood cells.

Then they started trying to tell surgeons what to do. It was normal surgical procedure to close wounds tightly as soon as possible; Wright's research showed they were better off keeping the wounds open to the air. It was normal procedure to dress the wounds tightly and change the dressings daily in an attempt to keep things dry;

Wright's research indicated that leucocytes thrived in moist wounds, which meant that the daily changing of bandages was not only exquisitely painful (the old bandages often adhered when they were pulled off) but increased the risk of infection. He recommended gauze soaked in salt water rather than dry bandages—bacteria hated salt, but leucocytes did fine if the concentration was right—plus, a damp salt dressing drew liquid from the wound, keeping fresh fluids flowing from the body, encouraging leucocytes and washing out bacteria. The old system of bandaging was all wrong; Wright's group came up with new methods and invented a new kind of bandage made from perforated celluloid to cover the gauze, allowing the wound to breathe.

And that was a critical point. Getting air to the wound was vital. While some wound-infecting germs (like strep) lived in the air, gas gangrene germs, the worst of the invaders, were known to be a species of anaerobe, bacteria that reacted to oxygen as if it were poison. Gas gangrene germs needed airless places to grow, while the body's tissues thrived on oxygen-rich air. If the goal was to prevent a gas gangrene infection, one of the worst things you could do was to cut off the air to a wound. It created an ideal growth environment for the bacteria.

And here was perhaps the most important thing Sir Almroth's group discovered: Wound infections proceeded in stages. First the new, open wound was colonized by aerobic bacteria, strep and staph, which grew quickly in a variety of tissues and loved oxygen. This first wave of infection would scavenge all the oxygen from a wound, clearing the way for anaerobes like gas gangrene germs, which then set up shop—especially in tightly closed wounds—and finished off the patient. Strep infections, his group discovered, were the underlying problem. They figured that 70 percent of the deaths from wound infections could be traced back to strep. If they could stop strep, they could stop wound infections. But they could not. For some unknown reason, no vaccine had ever worked against strep. And no drug existed that could stop a strep infection—or any bacterial infection—once it started in the body. Close the wound tightly and you risk gas gangrene. Leave it open and you invite in strep. The problem was more complex than anyone had thought.

Now that he had found what he thought was the key to wound infections—and found that he could not come up with a way to directly switch the process off—Wright focused on what could be done around the edges. He tried to get the surgeons to change their operating techniques. The surgeons' inclination was to leave tissue as intact as possible, then close as tightly as possible. But Wright's research showed that germs thrived in damaged or dead tissue; therefore it was necessary for surgeons to get more aggressive, cut out anything questionable, carve out a wound until it was free of hidden dirt and recesses, down to tissue you knew was healthy, down to the bone if necessary. Then keep the wound open for a while. Only after a few days, once it had clearly been shown that there was no infection, should a wound be closed to prevent any chance of further infection. "Close to sterilize," Wright wrote, "rather than sterilize to close."

He railed against the fast, wholesale evacuation of the wounded back to England. This, he felt, placed an undue strain on the patients during what might be the most important phase of recovery. The wounded needed to be kept in one place under proper treatment until the danger of infection passed, he argued; sending a patient with a tightly closed, dirty wound to England was often tantamount to sending him off to die.

Wright was very sure of his results and had no qualms about setting the surgeons straight as forcefully and directly as possible. At first the surgeons simply refused to listen. They were already overworked, thrown into the war, one day running a private practice in London where they might do a half dozen surgeries in a full afternoon, the next day in France cleaning and closing thirty or forty a day. The surgeons were proud of their ability to close with exquisitely small glovemaker's stitches; the best of them were proud of their ability to stitch below the top layer of skin to avoid scarring. The surgeons liked to leave the patient as intact as possible; as one surgeon said, "If I were to trim away all the badly damaged tissue in some cases, I should have to trim away half my patient." The surgeons believed in dousing everything inside and out in antiseptic. The administration wanted to ship the wounded back to England as soon as possible to clear desperately needed room for more.

It was all very depressing. With his research stalled, Sir Almroth wrote a friend, "I rather envy the people who are called to do something that is really worth doing, like dressing wounds or shooting Germans." Instead he spent his time trying to persuade the surgeons that what they were doing was more or less completely wrong. Domagk might have paid attention, but the skilled British surgeons at Boulogne were disinclined to listen to a twirler of test tubes. Even if they did listen, Wright's techniques would slow the necessary movement of patients through the hospital. The flood of wounded was too great to allow them to lie around for weeks with those salt-soaked gauze dressings. There simply were not enough beds in France. Everyone was being overwhelmed in a tidal wave of wounded.

It was not long before the Boulogne hospital administration requested that Wright himself be shipped home. But even if the surgeons were not interested in his findings, the Royal Army Medical Corps was. He was not transferred from his post, and his laboratory was kept intact. Eventually, as the war raged on and the thousands of wounded swelled into millions, his ideas began to have an effect.

The first change was the removal of more tissue in the operating room. Instead of saving every scrap of skin and tissue possible, surgeons switched to a new policy of cutting the skin back a quarter inch all the way around the wound, then exploring with the fingers—in Boulogne, at least, fingers that were often sheathed in sterile rubber gloves—cutting and scraping away all damaged tissue with a sharp, spoonlike curette, and finally scrubbing the wound thoroughly with gauze soaked in antiseptic. Wright believed that the first goal was to get as close as possible to "clean wound" in which bacterial levels were low enough to allow the body's own leucocytes to finish off the remainder. Given the limitations of antiseptics, this was the only approach that promised a cure. Researchers in World War I looked at bacterial infections like wars, in which competing armies tried to eradicate each other. If the patient won, the infection was over. If the bacteria won, the patient died. Wright's approach was to tip the scale toward the patient in as many ways as possible.

Wright welcomed the changes but still believed that surgeons tended to overuse antiseptics. The ultimate expression of surgical care during World War I was the "Dakin-Carrel Treatment." Parts of it took into account Sir Almroth's findings, parts did not. It was a three-step procedure: First open the wound wide and clean out all foreign bodies, dirt, and dead tissue; then flush and irrigate the wound constantly with a solution of bleach and boric acid (a pink liquid similar to commercial cleaning and laundering solutions) delivered via yards of rubber tubing inserted into various depths of the wound; and finally probe and sample the wound daily to track bacterial growth. It was a nightmare to administer—and to be administered to. The incisions seeped and oozed, the antiseptic burned; dressings had to be changed daily, and every change required resetting the tubes. Abdominal wounds were flushed with syringefuls of antiseptic every three hours, and the stumps of amputees were immersed in antiseptic for ten minutes every two hours, day and night, until bacterial counts fell to an acceptable level. The worst-off patients had to be bedded on rubber sheets. Even the bravest soldiers sometimes had to be strapped down. Dakin-Carrel did not stop the epidemic of wound infections, but it worked sometimes where nothing else could, prevented some secondary amputations, and saved many lives.

Still, in the final days of the war, as he weaved his way through the beds in the grand hall of the casino on his way to his laboratory, Sir Almroth was unhappy. The men sprouting tubes, the burning chemicals, the soldiers' anguished moans as their dressings were changed, was not what Wright wanted. He had wanted a cure—a fast, faultless, relatively painless answer to wound infections. He had worked through the war for four years, every day, virtually without a break. And he had failed. He missed his London home and his garden. He was fatigued—everyone in his group was exhausted by the war's end. Relations with his superiors were strained. He was "profoundly unhappy" during the war, one of his assistants remembered.

He understood that his work had had an effect. Nearly 2 million British soldiers were admitted to hospitals with wounds in World War I. One in five either died there or emerged a permanent invalid. Without his research the number of deaths would have been much higher. Still, when he returned home at the war's end, he found scarred and limbless men on every street, living reminders that the only ways still to stop infection came down to cutting and hoping. The big advances had been made in killing men, not saving them.

It all boiled down to a story that was told after the war of a veteran who had been gassed, coughing his lungs out in a hospital. He was unable to speak, so he scrawled a note to his brother: "This is what modern science has done for me."

CHAPTER THREE

After two years in the Ukraine, Domagk returned to Belgium as a medical aide, part of the troop buildup for the German Spring Offensive of 1918. Disease was everywhere. The long war, the hunger and cold, the crowded camps, the rotting carcasses, and the destruction of public health systems created an explosion of infectious disease. Out of it spread a great influenza pandemic. As he traveled west, Domagk remembered "cruel and shocking scenes," scores of "young and blossoming" soldiers dying around him, often before reaching the front lines, a mining camp he traveled through now a ghost town, miners dead before they could be hoisted from the shaft. No one seemed to know exactly what had caused the pandemic—some sort of bacteria, physicians supposed, but the pathologists (physicians who examined the bodies of the dead in great detail, searching for the causes and effects of disease) found such a mix of germs in victims' lungs that it was difficult to narrow the pandemic to a single culprit. Certainly no one knew how to stop it. As influenza raged around the world, the Spring Offensive failed, and Domagk's unit fell back. The Germans, Domagk included, thought their retreat was just another temporary setback in a trench war that had ground down to small gains and losses. But this time was different. Belgian flags were being raised on top of buildings. Civilians started shooting at the retreating soldiers from basement windows. Angry, starving mobs attacked the supply camps. Domagk helped chase townspeople away from the army's food, shooting flare pistols to scare them off. Rumors began: The German emperor had stepped down, there was a truce, the war was lost.

The war ended for Domagk while he was taking shelter with some comrades and civilians in the basement of a Belgian pharmacy. Official notification came that the emperor had abdicated. Germany had surrendered. Still Domagk's unit could not believe it. They were far from beaten. This was not a military defeat. Someone in government must have betrayed them. There was a growing sense of panic. There was talk of detonating explosives in the town, blowing up intersections. A wealthy Belgian tried to bribe Domagk's comrades with a packet of a thousand francs to keep them away from a street nearby where his mistress and child lived. They assured him that all civilians would be evacuated. Then the pharmacist in charge of the building opened his stock of Bourdeaux; Domagk remembered him telling the soldiers it was better for all of them, Germans included, to drink it now than to have a grenade blow it up later. A German junior physician started playing "Deutschland über Alles" on an old piano. They opened bottle after bottle. "What seemed unimaginable and incomprehensible to us just a few days ago, now, in spite of all the mad sacrifices, seemed senseless and without success," Domagk wrote. "The war lost!"

It was all confusing. His unit crossed the border back into Germany at Eupen, where the townspeople treated them like returning heroes, decorating the soldiers' cars with green garlands and little black, red, and white flags. They came home to a nation in political chaos. The victorious Allies occupied the industrial areas along the Rhine and began levying stupendous fines on Germany, reparations, they said, for the damage the Germans had wrought. The new German government put in place after the kaiser's abdication seemed shaky; the economy stalled; labor unrest and famine spread.

Domagk packed away his uniform, visited his parents, and retreated to his medical studies at Kiel University. Kiel was in the far north of Germany, toward Denmark near the Baltic. Here, as his nation struggled to regain its footing, he grew to love the area, the blue of the sea, the sky filled with light and crying gulls, so different from his home. He had inherited both a love of learning and a sense of discipline from his father, a schoolmaster, and he coupled those now with a natural skill at observation and the sense of mission he had found in the Ukraine. He became a model student. Germany's medical schools were the best in the world, and the rigorous program at Kiel was one of the best in Germany. Domagk's small student's room, packed with books, his desk decorated with a human skull, was often freezing cold. The postwar years in Germany were hard. Food was scarce. Domagk, cadaverously thin, once fainted from hunger and exhaustion during a physics lecture. He was not the only one.

Rather than become discouraged, though, he seemed inspired by the deprivations. He was a young man on a mission, dedicated to hunting down and destroying the bacteria that had killed so many of his comrades-in-arms. The war had immersed him in a swamp of disease and ignorance, chaos and blood. Medical school was a much-needed climb into clean air, the intricacies and mysteries of medical science offering a way back into a world more orderly, reasonable, and honorable. It was a way of not only fighting off the terror of war but of giving something back for all the lives lost. Hemingway called them the "Lost Generation," the ex-soldiers returned to a civilian life that made no sense, reacting with debauchery and despair. Domagk saw that happening, too, among his former comrades: "Many were not able to return to their work," he remembered. "Others were dancing on a volcano, some of them with broken wings. They were just brooding and under the perpetual influence of alcohol." But he refused to become lost. He decided that the best response to the war was measured thinking, cool logic, the careful application of science. He was not naïve. He understood that science had created terrible engines of war. But he also believed that science, properly used, could undo much of the damage men had done. Medical science in particular, he was convinced, held great promise. What other options were there? He believed that science at its best was a flower of Western culture, unbiased, apolitical, transnational, open, and progressive. It destroyed superstition and cant. It threw at least a little light into the darkness. And it worked. Domagk turned away from Lost Generation nihilism and, like thousands of others during the 1920s, joined the campaign for a better world through science.

Medical science was not an easy discipline to master. It required years of intense study, devotion, deprivation, memorization, immersion in mathematics, physics, chemistry, bacteriology, physiology, anatomy, pharmacology, and toxicology. He dove into his books, learning gradually to find his way through the tendoned and muscled, boned, blooded, and intricately balanced structure of the human body. Once he learned the basics, he applied what was known about the body's complex and often confusing relationships to disease. And in this way he eventually learned a great lesson of the age: With rare exceptions there was nothing he could do—nothing *any* physician could do—to stop a disease. Diseases could be *prevented*—this was a different thing, a matter of giving vaccines, getting people to wash their hands, cleaning up waterways, building better sewage systems, the triumphs of nineteenth-century public health—but once established in the body, once a disease had gained entry into a patient, the best-equipped and -trained physician in the world in 1920 could do little more to affect its progress than could a medicine man with a mask and bone rattle.

Or a medieval monk.

IN 1084, the devil visited Vienne. He took up residence in the French town's poorest hovels and richest villas, its markets and churches, and for a time he had his way. Those he touched shivered and fell weak, pain prodded their backs and heads, they grew hot and thirsty as if suffering the fires of hell, and then a rash burned their faces red. Sometimes their eyes swelled shut. Pustules erupted. Occasionally the rash spread to the limbs, rotting them somehow from within, veins of dark poison turning arms and legs black and stinking. How they suffered. Their skin grew so sensitive that many could not bear to be touched. Some in the medieval French city recovered. Many died.

God, too, however, was in Vienne, and performed miracles through the bones of St. Anthony, brought to the town by a local knight after a crusade. As the devil's affliction burned, the people of Vienne prayed to St. Anthony. And the saint answered: The signs of the devil subsided in the town, and then, as suddenly as it had come, the affliction disappeared. After that the burning rash, wherever it was seen, was called St. Anthony's Fire.

Banished for a time, it returned to Vienne eventually and visited other towns as well. St. Anthony's Fire flared and spread across Europe through the Middle Ages, taking one life in one town, a score in the next, sometimes attacking a single house, sometimes engulfing a village. Neither as horrific as the plague nor as deadly as cholera, St. Anthony's fire was yet another contagion in an age of contagion, part of the background of medicine, a constant threat. Smoldering steadily, century after century, it killed millions. Every physician knew it; none could cure it.

By the seventeenth century, enlightened physicians had stripped St. Anthony's Fire of its religious trappings (although their patients continued to pray to St. Anthony) and placed it firmly among the fevers, distempers, and rashes that arose from natural causes. The cause of St. Anthony's Fire—now becoming known under a more scientific name, erysipelas (from the Greek *erythros,* "red"; *pellas,* "skin")—was not the devil but, as physicians of the day understood, exposure to corrupted air, ill winds, putrescent vapors, miasma (*mal aria,* as the Italians called it), different names for the same thing: corrupted air rising from marshes, cesspools, or rotting corpses; vapors associated with fogs and cemeteries, swamps and storms, which, once inhaled, could attack and derange the four humors, the careful balance of which was critical to all human health. It was well known since the times of antiquity and a matter of medical doctrine that all human disease was rooted in unbalanced humors. Each person had within him or her the four, the melancholic, sanguine, phlegmatic, and choleric humors, embodied respectively by black bile, blood, phlegm, and yellow bile. Throw any of them out of kilter and illness could result. Therapy was based on rebalancing the humors through the common methods of bleeding, purging, and administering medicines related to the particular humor desired. This could sometimes be determined by matching the color or taste of the medicine to the humor in need of treatment. The miasma theory certainly explained why people in swampy areas or wandering about in chill fogs got sick more often. Because the reigning doctrine decreed that all diseases resulted from imbalances in the humors, all diseases were basically similar in nature: All fevers, for instance, were variations on a single great humoral Fever. The particulars of an individual patient's condition, the reasons one patient's fever appeared different from another's, could be attributed to specific factors in the environment, such as the positions of stars and planets.

At the time, corrupted air was as good an explanation as any for diseases like erysipelas. Of course, even in the seventeenth century some physicians recognized that in certain cases the patient did not appear to have been exposed to miasmas, had not taken a chill in a graveyard or traveled through a marsh, and here physicians thought that the humoral balance might have been upset by other factors: Violent passions, perhaps drinking to excess. Or staying too long in a warm bath. Or anything that overheated the blood.

In other words, they did not know what caused erysipelas, and without knowing what caused it, they had no effective way to stop it. Physicians of the seventeenth century might have been more scientific than the priests who treated the sick with wine washed over the bones of St. Anthony six centuries earlier, but the results were the same. Sometimes patients recovered, sometimes they died. Nothing the physicians did seemed to make an appreciable difference in the outcome.

That did not stop them from trying. For more than two centuries, they continued to use increasingly "heroic" measures—purging, bleeding, and dosing with chemicals—to balance the humors. In 1799, for instance, when the former president of the United States, George Washington, was treated for a severe throat infection, he received state-of-the-art medical care from the best physicians of the day. They gave him a compound of mercury (a poisonous heavy metal) by mouth and injection, administered a toxic white salt to induce sweating and vomiting, applied caustic poultices to blister his skin, asked him to inhale hot vinegar fumes that burned his throat, and bled him four times, taking a total of five pints. Then President Washington died. It is an open question whether his doctors might have saved his life by leaving him alone.

By Domagk's day physicians understood a great deal more about the human body and the effects of medicine. And the more they understood, the less they tinkered. The theory of the humors had finally been overthrown by new scientific discoveries detailing the nature of infections and, with greater and greater precision, the workings of the body. Medicine was now on its way to becoming a true science, with a firmer basis in biochemistry, a growing understanding of physiology, and treatment based on a degree of scientific proof rather than ancient dogma. By the 1920s the heroic age of medicine was long dead. The pendulum swung far the other way: The extreme treatments of the eighteenth century had been replaced with a conviction that it was usually better to do nothing, a deep conservatism rooted in the idea of strengthening the body rather than attacking it.

The new medicine Domagk learned was very complex, yet much of it could be boiled down to a simple guideline: When in doubt, leave the patient alone. Let the body heal itself. The more researchers learned about the human body—its marvelous machinery for repairing itself; its ability to shake off, in many cases, the worst diseases; its delicately balanced metabolism capable of maintaining temperature, salt levels, hormones all within very exact limits; the complex and highly effective defenses it mounted against invading microorganisms—the clearer it became that the doctor's most important jobs were to offer comfort and stand aside. Physicians lessened pain and relieved suffering; took away some fear by explaining to patients and their families what was happening and predicting what to expect next; provided good nursing, the "tenderest care," as one physician of the day put it; watched, waited, and hoped for the best. They had no tools with which to do anything else. Once an infectious disease started in the body, there were no drugs that could stop it (with the sole exception of quinine for malaria). In 1928, a pharmacologist estimated that of the thousand or so drugs available to physicians, only about one in ten was "by common consent deemed essential for treating the sick." Some considered that estimate too high. One contemporary physician wrote that there were only about a dozen drugs that worked reliably in treating disease in the 1920s, notably aspirin, insulin, quinine, digoxin for heart failure, and a few sedatives and painkillers. Beyond that there were unproven and often dangerous "patent" medicines and folk cures. Drugs were primarily sold directly to consumers, and quack remedies flourished. Good physicians developed a healthy mistrust of all drug claims, regardless of their source.

Because they could not cure, physicians had to be compassionate more than powerful, humanist more than scientist, care provider more than god. There was a basic humility among physicians in those days. "Whether you survived or not depended on the natural history of the disease itself," wrote the American physician Lewis Thomas, who earned his M.D. about the same time as Domagk. "Medicine made little or no difference."

Many physicians believed that the situation was unlikely to change. There was an underlying pessimism in medical schools about the chances of ever finding effective drugs, and there was an unwillingness to waste time trying. A physician doing drug research was a physician taken away from patient care. There was an unsavory aspect to a physician's developing a drug for money. There were ethical questions about testing drugs on patients. Developing new drug therapies smacked of a return to the discredited age of bleedings and purgings. New drugs in any case came mostly out of private firms and were sold with hyperbole at best and lies at worst; a physician's task was to sort through the dubious claims and push away the hucksters selling them. Medical education stressed understanding the body, accurately identifying diseases (the art of diagnosis), and providing supportive care. Young doctors in training did not need to be bothered with new drug theories. The men and women of medicine during that time were, as one historian put it, "therapeutic nihilists."

ANTONI VAN LEEUWENHOEK was a scientific superstar. The greats of Europe traveled from afar to see him and witness his wonders. It was not just the leading minds of the era—Descartes, Spinoza, Leibnitz, and Christopher Wren—but also royalty, the prince of Liechtenstein and Queen Mary, wife of William III of Orange. Peter the Great of Russia took van Leeuwenhoek for an afternoon sail on his yacht. Emperor Charles of Spain planned to visit as well but was prevented by a strong eastern storm.

It was nothing that the Dutch businessman had ever expected. He came from an unknown family, had scant education, earned no university degrees, never traveled far from Delft, and knew no language other than Dutch. At age twelve he had been apprenticed to a linen draper, learned the trade, then started his own business as a fabric merchant when he came of age, making ends meet by taking on additional work as a surveyor, wine assayer, and minor city official. He picked up a skill at lens grinding along the way, a sort of hobby he used to make magnifying glasses so he could better see the quality of the fabrics he bought and sold. At some point he got hold of a copy of *Micrographia,* a curious and very popular book by the British scientist Robert Hooke. Filled with illustrations, *Micrographia* showed what Hooke had seen through a novel instrument made of two properly ground and arranged lenses, called a "microscope." Hooke's device was simple and weak, something on the order of a child's toy today, but it was good enough to see previously invisible details of the structure of insects, bird feathers, cheeses, and sponges. Hooke's lenses disclosed a common flea "adorn'd with a curiously polished suite of sable Armour, neatly jointed," and a thin slice of cork "all perforated and porous, much like a Honey-comb" (this was the first description of plant "cells," a term Hooke coined). *Micrographia* was an international bestseller in its day. Samuel Pepys stayed up until 2:00 A.M. one night poring over it, then told his friends it was "the most ingenious book that I ever read in my life."

Van Leeuwenhoek, too, was fascinated. He tried making his own microscopes and, as it turned out, had talent as a lens grinder. His lenses were better than anyone's in Delft; better than any Hooke had access to; better, it seemed, than any in the world. The microscopes he made—small by today's standards, smaller than a hand—were far more powerful than Hooke's. With them he started looking at everything: bees' mouth parts, lice, fungi. He found that he could see more and much smaller things than Hooke had seen. He went beyond Hooke, turned his lenses on blood and spittle, paper and snow, linen, chalk, sugar, vinegar, tears, soap, bile, seeds, sweat, the liver of a sheep, the eye of a cow, hair from the tail of an elephant. He saw crystals form in evaporating salt water. He saw poison ooze from the sting of a scorpion.

Then, in the summer of 1675, he looked deep within a drop of water from a barrel outside and became the first human to see an entirely new world. In that drop he could make out a living menagerie of heretofore invisible animals darting, squirming, and spinning. He did not quite believe his eyes and for a time kept his notes to himself. At last, when he was sure, he published his findings, announcing the discovery of what he called extremely tiny "animalcules," "wee beasties," each with its own unique method of locomotion, what looked like tiny arms or fins waving, tails whipping, rotating, tumbling, zooming. They seemed to come in all sizes. The closer he looked, the finer the lenses he used, the smaller and smaller animalcules he saw. There seemed to be no end to them. It was then that the greats of the era began making their way to Delft to look through van Leeuwenhoek's magic lenses.

He kept looking at other things as well. He was interested in seeing, for instance, if something in the way spices were built could be related to their effect in the mouth. Perhaps, he thought, pepper might get its bite from little spikes on its surface. He softened some pepper in water to better study it and on April 24, 1676, turned the best of his microscopes to the pepper water. There he found, at the very limits of his vision, what seemed to be living things far tinier than any he had ever seen before, so infinitesimal that he judged that "thirty millions of these animalcules do not cover as much space as a coarse sandgrain." It was humanity's first look at bacteria. Using his best lenses, he began finding them everywhere—in dirt, in water fresh and salt. The scrapings from his teeth were a particularly rich source. He hired a local artist to draw what he saw and sent his findings to the greatest scientific body of the day, the Royal Society in London.

Van Leeuwenhoek's raising of the curtain on a new world was greeted with what might kindly be called a degree of skepticism. Three centuries later a twentieth-century wit wrote a lampoon of what the Royal Society's secretary might well have responded:

Dear Mr. Anthony van Leeuwenhoek,

Your letter of October 10th has been received here with amusement. Your account of myriad "little animals" seen swimming in rainwater, with the aid of your so-called "microscope," caused the members of the society considerable merriment when read at our most recent meeting. Your novel descriptions of the sundry anatomies and occupations of these invisible creatures led one member to imagine that your "rainwater" might have contained an ample portion of distilled spirits—imbibed by the investigator. Another member raised a glass of clear water and exclaimed, "Behold, the Africk of Leeuwenhoek." For myself, I withhold judgment as to the sobriety of your observations and the veracity of your instrument. However, a vote having been taken among the members—accompanied, I regret to inform you, by considerable giggling—it has been decided not to publish your communication in the Proceedings of this esteemed society. However, all here wish your "little animals" health, prodigality and good husbandry by their ingenious "discoverer."

The satire was not far from the truth. Although very interested in the Dutchman's discoveries, so many English scientists were doubtful about his reports that van Leeuwenhoek had to enlist an English vicar and several jurists to attest to his findings. Then Hooke himself confirmed them. All doubt was dispelled. The invisible world and its myriad minuscule denizens were real. Several generations of scientists spent the next two hundred years sorting and classifying the microorganisms, trying to see what it all meant and how it fit with the rest of the world—the visible, known world. Some of the early work was, in retrospect, amusing. Dedicated researchers, their eyes blurry from looking through primitive microscopes, thought they saw lips on the microbes, tiny eyes or mouths. Some believed there was only a single type of bacterium that changed depending on its environment.

For a long time, this meant little to medicine. The problem was that bacteria—the pepper-water animalcules, the very smallest beasts van Leeuwenhoek could see—lived profligately, in huge, chaotically mixed bunches, neighborhoods and cities and nations of bacteria. Look at the scrapings from human teeth mixed with a bit of water and even van Leeuwenhoek could make out a swarm of one or two dozen different types of bacteria (today, with better equipment, we know that hundreds of different types of bacteria can be found in the human mouth), all mixed together with dead cells and unclassifiable flotsam and jetsam. It was not until the 1870s that a French chemist named Louis Pasteur did some ingenious work linking the invisible world to disease, demonstrating that both fermentation—the making of wine and beer—and putrefaction—the rotting of meat—were due to the action of yeasts and bacteria. If bacteria could rot meat, Pasteur reasoned, they could cause diseases, and he spent years proving the point. Two major problems hindered the acceptance of his work within the medical community: First, Pasteur, regardless of his ingenuity, was a brewing chemist, not a physician, so what could he possibly know about disease? And second, his work was both incomplete and imprecise. He had inferred that bacteria caused disease, but it was impossible for him to definitively prove the point. In order to prove that a type of bacterium could cause a specific disease, precisely and to the satisfaction of the scientific world, it would be necessary to isolate that one type of bacterium for study, to create a pure culture, and then test the disease-causing abilities of this pure culture.

In theory all that was needed was to find a way to separate a single bacterium out of the mass, isolate it, feed it, and allow it to multiply (no sex necessary; most bacteria simply divide to multiply). Growing bacteria is easy: Anyone can leave a flask of warm beef broth open on a table and return a day or two later to find it cloudy with billions of bacteria. But it would be a chaotic mix of types. To study bacteria it would be necessary to isolate a single one and let it grow into a pure culture. Selecting out one kind of bacterium with the tools available in Pasteur's day, however, was like trying to pick up a single grain of sand with a steamshovel. So Pasteur's idea—his "germ theory" that claimed infectious diseases were caused by bacteria—remained just one of many explanations of the source of infectious disease in medicine, along with "crowd poisoning" (disease spread by emanations of respiration), diseases caused by effluvia from the skin, spontaneous generation of worms and fungi, decomposing excreta, and diminished oxygen. Before it could be proven that bacteria caused disease, it would be necessary to find some way to study them in pure culture.

The man who made that possible was Robert Koch, a small, nearsighted German physician with a rural practice in Prussia. For his twenty-eighth birthday in 1871, Koch's wife bought him a microscope, and, like van Leeuwenhoek, Koch started looking at everything. In drops of blood taken from sheep and cows that had died of anthrax, he saw what he thought might be bacteria: threads and rods that he could not find in the blood of healthy animals. Of course he had read about Pasteur and the idea that bacteria might cause disease, but Pasteur's germ theory was just that: a theory. Now as Koch began thinking about the problem, it fell into two parts: First, how could he determine if these threads and rods he was seeing were actually bacteria and not some bits of broken-up detritus in the blood; and second, if they were bacteria, how could he prove that they were causing anthrax? He was a small-town doctor. He knew only what he had been taught in medical school (which included next to nothing about bacteria) and what he'd read. He had no equipment. So he improvised. He did not have syringes, so he sterilized slivers of wood by heating them in the oven, dipped the slivers into the blood of anthrax-infected sheep and cows, and pricked the bacteria into healthy mice. The mice died. He opened up their bodies and examined the organs: They had the same signs of infection, the same swollen, blackened spleens found in anthrax-infected sheep and cows. Their blood was swarming with threads and rods. He then used his splinters to transfer blood from the dead mice to healthy ones, again passing the disease. Still, this was not definitive proof that bacteria caused anthrax. The blood he was using, even a tiny drop, was a mixture of thousands of bits and pieces of cells, what might be other bacteria, what might be motes dancing before his eyes. It seemed logical that his rods and threads were alive and multiplying and causing anthrax, but this was no way to prove it. He needed to separate the suspected bacteria from whatever else was in the blood. He needed a pure culture.

The solution came through the eye of an ox. From its interior he drew a few drops of clear liquid that he figured might make a good growth medium for bacteria. He examined the liquid closely under his microscope and found the field clear and empty, no bacteria that he could see. It was sterile. Then he carefully added to this growth medium tiny bits of spleen from a mouse that had been infected with anthrax. After a few days, he found the eye liquid teeming with rods and threads. Dead cellular garbage would not grow like this; it appeared that his rods and threads were live bacteria. He then used the bacteria-rich ox-eye liquid to infect other animals, finding that rods and threads transferred on the end of a splinter jabbed into a test animal would kill it in a few days, the dead body swarming with billions more rods and threads. Koch studied the bacteria and became an anthrax specialist, the first to see the rods and threads produce spores, the first to postulate ideas about anthrax transmission, forgetting now, mesmerized by his microscopic world, about his local patients. He believed he had proven that bacteria—a single type of bacterium—caused a specific disease. Regardless of the fact that he had no standing in the field of bacteriology, he began publishing his findings. His articles, carefully written, logically presented, began to convince others.

One germ = one disease was a simple and powerful concept but, even after Koch's first discoveries, a difficult one to prove definitively until better ways could be found to create pure cultures of more types of bacteria. Early bacteriologists designed one Rube Goldberg contraption after another to separate and purify bacteria. Many depended on taking a very small sample of infected material, as Koch had, and growing it in liquid, then taking a very small sample of that and growing it again, repeating the process until just one type of bacterium could be found. Koch had managed to do it. But others found it nearly impossible. No matter how ingenious the machinery, how careful the researchers, they kept ending up with beakers of mixed bacteria. The inability to get anything but mixed cultures led many scientists to believe that bacteria had to be in mixed groups in order to thrive, that they could never be separated, perhaps that they were capable of changing from one type into another.

Then came Koch's potato. The country doctor, his achievements already recognized and rewarded by appointment to a prestigious position in Berlin, able to hire assistants and run a large research facility, stumbled across the answer to his research questions in the form of the cut half of a boiled potato. It was sitting forgotten on a table. Just the kind of thing to bring contamination into the lab. He picked it up to throw it away. Then he noticed a sprinkling of dots of various sizes and colors on the cut surface. Curious about what was growing, he used a bit of wire to pick a smidgen from one of the dots and examined it under the microscope. He saw bacteria, multitudes of them. But it was not a mix. They were all identical. He cleaned his equipment, picked a bit from a different dot, and looked again: again bacteria, and all again identical. Each dot, he recognized, was a pure culture. Then he realized what was happening. Single bacteria in the air were drifting onto the surface of the potato, sticking in place, feeding and multiplying, the one dividing into two, the two into four, four to eight, creating a crowd of direct descendants, a growing dot, a colony, each colony descended from a single bacterium. *Each dot a pure culture.* The trick was using solid food instead of a liquid. He and his assistants quickly explored the idea by gently touching the ends of sterilized wires into mixes of bacteria and then streaking them out on a plate layered with food. They needed a flat surface that grew bacteria easily, so they experimented with gelatins blended with meat broths and nutrient mixes, a sort of kitchen cookery in which the mixtures were heated until they were sterile, then poured into plates to cool. Across the solid gelatin, they streaked a wire dipped into bacteria, first one way, then another, each time separating and spreading the bacteria more, getting to the point where they were pulling a few individual bacteria across the surface, each of which would grow into an individual colony, each a pure culture. One of Koch's students, R. J. Petri, designed a shallow glass plate with a removable top to keep out contaminating bacteria, the first petri dish. Another in Koch's lab found the ideal growth medium, a dried seaweed extract from Asia that could be dissolved and heated together with food like sterilized whole blood or meat broths, then poured and cooled to form a solid base. The Malayan name for the seaweed was agar-agar. For the next century, agar gelatins laced with nutrients in petri dishes would become the most important tools available to bacteriologists. Koch's techniques were simple and revolutionary. Bacteria grown in pure culture on his plates could be studied one by one, examined and sized under the microscope, characteristic shapes noted, growth rates charted, preferred diets found. And, finally, they could be linked definitively to human disease. Koch came up with a set of rules for proving the link: You first had to find the germ present in all cases of a given disease; you had to sample it from a disease victim and grow it in pure culture; you had to use the pure culture to cause the disease in a test animal (human or otherwise); you then had to isolate the same germ again from the diseased test animal; and finally you had to demonstrate that you ended up with the same type of germ you started with.

Using Koch's Postulates, as they became known, scientists during the 1880s and 1890s searched out and identified the bacterial causes of one infectious disease after another: diphtheria, tuberculosis, anthrax, pneumonia, tetanus, cholera. One germ = one disease.

THIRTY YEARS after Koch's discoveries, when Domagk was in medical school, the bacterial cause of many infectious diseases was accepted as fact. Domagk learned how some bacteria released poisons called toxins that could weaken, even kill, patients; how others invaded tissues and organs; how some thrived in oxygen-rich blood, while others thrived in oxygen-free places such as the intestines. Like van Leeuwenhoek, Koch, Pasteur, and a host of others, he became fascinated by the microscope and the invisible world it revealed, the way you could *see* the microorganisms going about their lives, the way the tissues in the body were affected by them, the ways the body fought back, tried to rid itself of infection. Here at the microscopic level was the drama of infection, the parts played by the bacterial killers that had destroyed his friends during the war. Here, deep in the body, had to be the secrets to defeating them. He learned everything he could about bacteria and about life at the bacterium's level, the microscopic level, the chemical level. His professors noted his tendency toward the laboratory and away from dealing with people. Not that he was bad with people—he was liked well enough, although he seemed to have no close friends. He showed signs of ambition and a real talent for laboratory work. But he was somewhat distant, a bit sad. There was that skull on his desk. He kept his distance, a reserved young man who seemed to come alive only when talking about the causes and cures of disease.

In 1921, Domagk, age twenty-six, graduated from medical school. He received a *sehr gut,* the highest possible mark, for his doctoral dissertation on the biochemistry of muscle cells. He was now in a position to begin doing something about the madness he had seen at the field hospital. He was now a physician.

CHAPTER FOUR

DOMAGK'S FIRST POST was at Kiel's main city hospital, not far from the university. Here he began training as a novice doctor under Georg Hoppe-Seyler, an expert in internal disease, a man from whom Domagk had taken classes, and a teacher he admired. While it was a good place to see infectious diseases in action, the hospital was not well equipped for tracing them to their roots—its laboratory support, access to microscopes, and chemistry equipment were minimal. Still, Domagk was, for the moment, satisfied. He was earning a bit of money at last and had at least a few hours of free time for concerts, art exhibits, and theater. He enjoyed exploring Kiel, a Baltic port city and home to Germany's largest naval base. He loved looking at the sea. And he found a good friend in another young doctor just getting started, Ferdinand Hoff. They shared an interest in infectious disease and spent hours talking about how bacteria spread disease and how the body tried to fight it off. Hoff was especially keen on the role of the immune system, an enthusiasm he passed on to Domagk. The hospital had wards filled with lung cases—tuberculosis and pneumonia—in which Domagk took a special interest. Tuberculosis (then often called consumption in English-speaking countries) was the greatest killer in Europe. It had become almost a cliché in novels and plays—an invisible villain in *The Magic Mountain* and *Camille*—as characters coughed, looked at their handkerchiefs, and saw spots of blood. TB could attack many areas of the body, but the lungs were the main target. There it set up a slow, inexorable infection that turned healthy lung tissue into something with the consistency of soft cheese (the doctors called this display of the disease caseation necrosis, from *caseus,* Latin for "cheese"). Koch had shown that TB was caused by an unusually hard-to-grow and hard-to-study bacterium covered in a waxy coating. The disease resisted almost every form of treatment. All you could do was provide supportive care, good food, and rest. And hope. As always, regardless of what doctors did, some patients recovered while others died. Why? Hoff and Domagk talked about it and came to believe that the difference lay inside the body, at the level of its own defenses, within the immune system. But where? What was it within the immune system that made the difference? They began looking, turning their microscopes to the known centers of the immune system, the patients' livers, spleens, and blood.

"*Das Blut ist ein ganz besonderer Saft*" (The blood is an absolutely remarkable fluid) Goethe wrote in *Faust.* The German focus on blood as a combination of mythic symbol, political ideology, and biological mystery was long-standing. Domagk saw the truth of it under his microscope: Human blood to the naked eye was a red liquid, but magnify it enough and it became a wonderland. Properly stained with chemical dyes to bring out its details, blood revealed a menagerie of cells suspended in a clear liquid: red blood cells—simple bags of hemoglobin, the chemical that carried oxygenated blood—and white cells, mediators of the immune response. The more white cells were studied, the more complex their tribe became, including various monocytes, granulocytes, lymphocytes, and macrophages. The white cells, Domagk had learned in medical school, were what mattered when it came to fighting disease. Some—including Sir Almroth Wright's ultimate antiseptic, the leucocytes—were predators, roaming the body seeking out and engulfing invaders. Others were like little factories, pumping out chemicals called antibodies that were somehow able to distinguish among different kinds of bacteria and attack those threatening the body. The immune system was an intricately complex system of cells, organs, chemical triggers, and messengers. No one at the time Domagk was in school understood what the parts all were, how they all worked, or how the different parts coordinated to fight infection. No one understood why the system sometimes failed. But it generally did work, and when it worked, it worked beautifully. A strong immune system could throw off even a severe infection; a weakened immune system in an old person or a severely ill patient could be overwhelmed by what appeared to be a mild disease. There were no simple answers. So Domagk drew blood, examined it under a microscope, put it in test tubes, and spun it in a centrifuge until the blood separated into layers, the bottom half dark red, packed with the heavier red blood cells, the upper half a clearish, straw-colored liquid. The clear liquid was called serum. It was a miraculous substance. Even with no cells in it, serum still contained the brew of mysterious chemicals that white blood cells released into the blood, including antibodies. Antibodies worked like magic. They were exquisitely sensitive. These chemicals were somehow able to distinguish between the body and those things that were not of the body, invaders like bacteria. They could distinguish between one kind of bacterium and another. They could, it was just being learned, distinguish between two chemicals that differed by only a few atoms. No one knew how.

Researchers did know, however, that serum could be used as medicine. Serum therapy was all the rage when Domagk was at Kiel. The idea had started back in the eighteenth century when Edward Jenner noticed that English mikmaids infected with a relatively mild disease called cowpox seemed not to contract the fatal disease smallpox. Without any idea of why it might work, Jenner—in a manner that would probably be considered criminal today—inoculated several poor rural children with the pus of cowpox scabs, then exposed them to smallpox. The inoculated youngsters were far more resistant to the disease than were children who had not received the cowpox. It was the beginning of the practice of vaccination: the exposure of the body to a suspension of less dangerous, weakened or dead microorganisms in order to jump-start an immune response. The body somehow "remembered" the vaccination for years and, when a real infection started, was able to fight it off much more effectively. The practice had been brought to a fine art by researchers like Sir Almroth Wright, who knew that when properly treated a patient's immune system could often fight off even the most severe infections. Confusingly, some types of vaccinations worked beautifully, such as Wright's typhoid vaccine, while others failed entirely—like his wound infection vaccines. No one knew why. The problem was that the immune system was still a complete mystery in many ways—especially concerning the nature of antibodies, those chemicals produced by some types of white blood cells within the body that were perfectly targeted to specific invaders.

Domagk's generation was just beginning to pull apart and examine the system. The immune system as they knew it had two ways of operating: the cellular response and the humoral response. The first consisted of a family of hunter-predator white blood cells like leucocytes that engulfed invaders and destroyed them—literally ate them in a process called phagocytosis (from the Greek *phageîn,* "to eat"). The second, the humoral response, consisted of an array of chemicals including the antibodies that circulated in the serum, substances that somehow also worked together and with white blood cells to help clear out invaders.

There was much still to learn about the humoral response. None of the chemicals were well understood. A worldwide research effort was under way to learn more. What Domagk understood was that by inoculating a test animal—that is, introducing a foreign substance under the skin and into the animal's body, shooting a particular germ or part of a germ, for instance, into a rabbit, a mouse, or a guinea pig—giving the animal a week or two to react, then drawing its blood and spinning it down, a researcher would find the animal's serum full of antibodies *targeted specifically to that particular germ.* Collect the animal's serum and shoot it into a human and it acted like medicine. The antibodies made by the test animal would attack disease in the human, but only if the bacterium that had been used to inoculate the animal was identical to that infecting the patient receiving the serum. The great German researchers Emil von Behring and Paul Ehrlich, students of Koch's, cured thousands of cases of diphtheria using the technique. The demand for their antidiphtheria serum grew so great that they used horses to make the serum in batch lots. For a while in the 1890s, it looked as if serum therapy might be the answer to all human disease.

But there were serious problems. Serum therapy in practice turned out to be difficult, costly, and sometimes dangerous. It was almost too precise. Each serum had to be matched exactly to the specific bacterium involved in the disease, which meant that doctors had to first diagnose the disease with great accuracy, then isolate the invading bacterium, grow it, identify it—there were, for example, twenty-odd strains of pneumococcus that caused pneumonia, each strain requiring its own serum—then find or create a serum to match. By that time the patient might be dead. In any case, having to make hundreds, even thousands of designer sera, each targeted to a single one of thousands of strains of disease-causing bacteria, made the process very expensive. Von Behring and Ehrlich's antidiphtheria serum worked as well as it did mainly because it was directed not at the bacterium but at a poison the bacterium produced, a poison that was identical in all diphtheria cases. Serum therapy had side effects, too; the human patient's immune system could view the injected animal serum as yet another foreign invader and mount a defense against the medicine, setting off a cascade of misguided reactions in the blood—serum sickness, they called it, a sort of allergy to the medicine—which could sometimes be fatal. Finally there was the ultimate mystery: Just as with vaccines, some diseases responded well to serum therapy, while it was powerless to stop others. Sometimes the magic worked, sometimes it did not. No one knew exactly why.

Still, for a generation, serum therapy was the only thing going. In the United States, millions of dollars were spent developing antipneumonia sera for every strain they could find, a brave effort that worked well in only one place, New York City, because only there were physicians well equipped enough to type the bacteria quickly and the city wealthy enough to stockpile a full array of sera. Sir Almroth Wright had boasted before the war that "the physician of the future will be an inoculator." In the 1920s many researchers continued to believe—especially in the absence of any effective alternatives—that the immune system was the only answer to infectious disease.

The more Domagk learned about the immune system, the more complex it seemed to become. Researchers throughout the world kept discovering more subtypes of immune cells, more chemicals in serum, more interactions between cells, more types of reactions within the body. Domagk, like many others, maintained faith in the body's ability to fight off disease. In all this immune mystery and confusion,

Domagk thought, there was the chance to find something new, something that could be applied directly to human disease. So he spent his time learning how to inoculate animals, how to dissect and examine the organs involved in the immune response, how to search through the catalog of the blood.

It was becoming ever clearer that Domagk was better suited to the laboratory than to treating patients. Recognizing that his mission was to find cures for many rather than healing for a few, he decided to specialize in pathology—the work of physicians who search for the root causes of disease, "that precinct of medicine that is often said to be the best suited to scientific (and dispassionate) intellects," as one medical historian put it. The word's root is the same as that of "pathos," the quality that arouses pity or sadness; it comes from the Greek word for suffering. In that simple sense, it was the perfect field for Domagk's life's work: Pathology is the study of suffering. It was also a well-respected medical specialty in Germany; there were career paths here involving many of the skills Domagk enjoyed, especially looking at tissue slices and smears under the microscope, tracking the effects of disease on organs and tissues. Pathology was entering a golden age as scientists armed with ever-more-powerful microscopes and biochemical tools started teasing apart the dizzying array of cells and chemicals inside the living body. Always eager to learn the latest techniques and findings, Domagk attended a 1923 meeting of the German Society for Pathology in Leipzig, where he met his next mentor, Walter Gross, the relatively young director of the Institute of Pathology at the University of Greifswald. Gross's work impressed Domagk, and the two spoke. Gross held out the possibility of working together. Kiel was not really the right place for Domagk long-term, but at Greifswald (another Baltic town about a hundred miles east) he would have a fully equipped university laboratory and the chance for real research. The university itself was old and respected. When Gross offered him a job as a junior doctor, Domagk eagerly accepted. The position would give him a sympathetic boss, access to better equipment, the chance to publish, and a bit more financial stability.

With his position ensured, he was now prepared to start a family. In July 1923, at age twenty-seven, Domagk decided to make his feelings known to the village girl in the white Communion gown, the sweetheart whose photo he carried when he was shot. Her name was Gertrude Streube. They had stayed in touch by mail for the nine years since he had set off to war, a childhood romance delayed by his insistence on a decent income before marriage. Gertrude in the meantime had provided for her own future, taking a well-paying job with the German Chamber of Commerce in Geneva. Domagk arranged to meet her at Lake Constance, a beautiful spot where he intended to propose. When the big day came, he went to the Greifswald station and found the train he wanted filled, so he waited for the next. Even the backup was almost full; he had to take a seat far in the rear. Before his train got to Munich, it was delayed by some sort of mechanical problem outside of Kreiensen. Domagk decided to get something to drink while the train was stalled. He got off. A few moments later, he heard a terrible crash. The train behind them, unaware of the delay, had smashed into Domagk's and demolished the car he had just left. Forty-eight people were killed. It was Germany's worst train accident in forty years.

Gerhard Domagk was lucky. First there was the bullet that had knocked off his helmet but missed his brain. Now the train wreck. And the four-leaf clovers . . . When he was a child, he had roamed the fields with his mother, gathering crimson poppies and blue cornflowers that grew near their home in the lake country. He became an expert at finding four-leaf clovers, his keen eyes able to see what no one else could. Sometimes, before the end of the day, he had gathered a small bunch. He amazed his colleagues later in life by spotting four-leaf clovers as they walked by a field, able to pick them out even from a moving bicycle, finding several where no one else could find one. In the Bayer Archive, among his piles of correspondence and lists of prizes, a four-leaf clover still lies pressed between two sheets of paper. He had sharp eyes and a bit of luck—good qualities for a scientist.

LUGGAGE DESTROYED, romance derailed, Domagk made new arrangements to meet Gertrude for Christmas 1923 in Dresden. There, in one of the beautiful city's most beautiful buildings, the Gemäldegalerie, standing before a Raphael of a Madonna descending from the clouds tenderly holding the Christ Child, he proposed marriage. She accepted. They could not marry immediately. Arrangements had to be made, money saved, Gertrude had to quit her job and relocate. Nearly two years would pass before the ceremony. But there was a part of Domagk that preferred that all things be in order—perhaps a reaction to the war—and things were in order now.

He began work at Greifswald in earnest. It was a small, pretty university town. He got along very well with his new boss, Gross. Domagk was able to focus on his cells and dyes for much longer and without the distraction of patients. Looking through his microscope, he saw something that he believed had never been reported: some star-shaped cells in the liver called Kupffer's cells that he observed attacking and eating—phagocytosing—red blood cells, engulfing and tearing them apart, spitting out the iron-containing remains of the hemoglobin. Domagk thought he had discovered something important, but a library search quickly showed that phagocytosis by Kupffer's cells had been reported fifteen years earlier. "I was not sad, on the contrary!" Domagk wrote. "My self-confidence grew and I told myself, 'If you have discovered such important things by yourself, just as other good observers have done before you, then you as well will discover something groundbreaking.' " He turned his attention to the larger system in which Kupffer's cells played a part, a vaguely understood setup designed, it appeared, to clean out old, dying red blood cells in the body so they would not clog up blood vessels. This was associated with the immune system, too. It included phagocytic cells like the ones he had seen in the liver, and it played a supporting role in fighting infections. Kupffer's cells and others in this system were intensely phagocytic, sitting in their microscopic tissue caves waiting for prey, able somehow to recognize the right targets, grab them, dissolve them, and destroy them. The targets included bacteria as well as old red blood cells. They were also good at ingesting dyes, which made them easier to study under the microscope.

Perhaps in this little-studied area was some factor, some chemical, that might be useful to medicine. Domagk focused his attention on Kupffer's cells. He gathered blood and tissue samples, fixed and stained slides, and spent hours staring into his microscope. He figured these cells might be secreting some chemicals to help them do their work, perhaps some undiscovered substance that would help fight human infections. But to study any new chemicals he found, he would need to do tests in animals, infect them with bacteria, then see if they were helped by anything given off by his phagocytosing cells. Animal tests were crucial in turning medicine from an art into a science. With mice or rabbits, you could do experiments impossible to do on humans. You could infect a number of them with deadly bacteria, give some of them the chemical you thought might work as a medicine, keep other infected animals aside as a "control" group (whatever the experimental substance was, controls did not receive any), vary experimental treatments at will. He started using mice. That was half the setup; now he needed the other half, a strain of bacteria that could kill the mice quickly and without fail—a certain end-point for the experimental subjects—and preferably one that was also a disease threat for humans. He started with a strain of *Staphylococcus,* a common germ in hospitals, cause of a number of human infections, and one that reliably killed unprotected mice. Using his mouse/staph system, he tracked how the animals responded to an infection. Shoot staph into a mouse, he found, and you could find the bacteria soon afterward picked up by some of his phagocytic cells and destroyed. In the process the cells released a breakdown product called amyloid, a gooey protein. Domagk next started trying various ways to boost the power of the animals' response. After months of work, he demonstrated that if the animals were "sensitized" beforehand to the staph—inoculated, in other words, with a small dose of dead bacteria, for instance, to jump-start the immune response before being injected with live bacteria—his cells seemed stronger, able to clear the invaders faster. That was interesting but expected: Boosting the immune response was what inoculations were all about. But he also found something else. If he injured the live staph before injecting them, by exposing them, for instance, to a weak antiseptic—not enough to kill the infecting organisms, just weaken them—his cells cleared them more efficiently. If a substance could be found—not an antiseptic; antiseptics were too harsh to use inside the human body; they killed normal body cells just as well as they killed bacteria—something new, something milder, something that could be injected into the body and would weaken the bacteria, then the body would have a better chance of cleaning up the rest. It would tip the balance in the body's favor. Not a harsh antiseptic that would injure the patient. Not a killer drug. An injuring drug.

THE SEARCH FOR ways to rid the body of dangerous bacteria started almost as soon as bacteria were suspected to cause disease. After Koch and Pasteur, physicians knew what bacteria were and knew that they caused diseases. The only thing they did not know was how to stop them.

And this brings us to Queen Victoria's armpit. In 1871, Her Royal Majesty, queen of the United Kingdom of Great Britain and Ireland, began suffering from a painful abscess, a boil, under her arm. Boils were more than a nuisance; they were bacterial infections that could, if left too long, lead to dangerous blood infections. Her Majesty's personal physician determined that an operation was needed to drain the infection. The problem was that the majority of operations in those days, no matter how minor, no matter how carefully done, resulted in postoperative infections that were often worse than the original condition. Everything got infected in those days before Koch, because surgeons did not understand what caused infections. The badge of honor for a hardworking surgeon (whose motto was "Cut through everything soft, saw through everything hard, and tie everything that bleeds") was a bloodstained frock coat worn in the operating room, the stiffer with gore the better. No attention was paid to germs. Bacteria causing disease was a theory in those days, not a fact. So surgeons picked up their instruments with bare hands and set them out on any handy table. They wore no masks. Because postoperative infections almost always happened, instead of trying to prevent them, they graded them. One sign of success was considered to be the appearance of "laudable pus," creamy white exudate of the wound, indicating a strong body and harbinger of a proper cure. In reality, of course, laudable pus and all other forms were signals of a potentially dangerous bacterial infection.

Postoperative infections carried off thousands of patients and scared thousands more away from the hospital. As a Scottish physician of the day put it, "The man laid on the operating table . . . is exposed to more chances of death than the English soldier on the field of Waterloo."

Then came Joseph Lister. He was a consummately skilled British surgeon who was devastated when he found that all his skill seemed to make little difference. He despaired as patient after patient succumbed a few days after his best work to "hospital fever," another name for postoperative wound infection. In Lister's early years, the mid-1800s, half of all amputation patients died from hospital fever; in some hospitals the rate was as high as 80 percent. Lister, like all surgeons, had little idea of how to improve the situation. Then he chanced on a newspaper article that caught his interest. It described how the residents of a local town, tired of the smell of their sewage, had begun treating it by pouring into their system something called German Creosote, a by-product of coal tar. Something in the creosote stopped the smell. Lister had heard about the work of Pasteur, and he made the same mental connection the French chemist had: The stink of sewage came from putrefaction, rotting organic matter; the stink of infected wounds also came from putrefaction; whatever stopped the putrefaction of sewage might also stop the putrefaction of infected wounds. So Lister decided to try coal-tar chemicals on his patients. And he found one that worked exceptionally well: carbolic acid, a solution of what today is called phenol. He became convinced, even as Koch was confirming the point, that bacteria caused wound infections; he further understood that carbolic acid killed bacteria. Over several years Lister expanded his use of carbolic acid; soon he was washing everything in the operating room—hands, scalpels, dressings—wiping tables, scrubbing incision areas, cleaning the catgut he used to sew up his patients. It worked, at least to a point. The rate of postoperative infections among his patients was dropping, although he never seemed able to stop them entirely. So he decided to permeate the entire operating theater with carbolic acid. He fitted a spray bottle, the type used for misting perfume, to a mechanism manned by an attendant turning a crank, filling the air in the operating room with the sharp chemical smell of carbolic acid. Lister's insistence on stopping the transfer of bacteria in the operating room became absolute. Once when a visiting knighted physician from King's College idly poked a forefinger into a patient's incision during one of Lister's operations, Lister flung him bodily from the room. By now, although he still lost some patients to hospital fever, he was losing far fewer than anyone else. Lister earned a reputation as the world's safest surgeon.

It was no wonder he was called to attend to Queen Victoria's abscess. Regrettably, during the procedure to drain the boil under her arm, Her Highness accidentally received a puff of carbolic acid full in the face from Lister's spray apparatus. She turned her wrath on the assistant at the crank. "I am only the man who works the bellows," he muttered miserably. That aside, the operation was a complete success. Lister became the first Englishman ever to be made a lord for medicine. And a new vocabulary became common in medicine: An infected wound was a "septic" wound (from the Latin for putrefaction); chemical cleansers like carbolic acid that could effectively kill bacteria were "antiseptics" or "disinfectants." A clean operating room was a "sterile" operating room, one free from living microorganisms. Antiseptic surgery using sterilized instruments very quickly became the norm. Lister revolutionized medicine.

The next three generations of caregivers practiced "Listerism" as avidly as previous generations had practiced bleeding and purging. Some physicians went antiseptic crazy. They ordered patients to inhale creosote vapors and gargle with mercurochrome. They injected antiseptics into their patients' blood, looking for a way to achieve "internal antisepsis," clearing bacteria out of the blood. Unfortunately, antiseptics like carbolic acid were strong, indiscriminate chemicals that killed healthy human cells just as quickly as they killed bacteria. Internal antisepsis was impossible with the chemicals they had—and it is likely that a number of patients were hastened to their end by the overenthusiasm of their doctors. By the time World War I began, the greatest excesses had given way to more rational use of antiseptics. These harsh chemicals were reserved for external use only. Properly used in this way, it was believed, antiseptics would make all hospital fevers—all wound infections, including gas gangrene—little more than a footnote in medical history. The British medical experience during the Boer War seemed to prove the point.

Then came the disastrous early months in Flanders, with their waves of gangrene and other infections. It seemed all of Lister's lessons were somehow proving false. At Boulogne, the model hospital that included Sir Almroth Wright's lab, Lord Lister's ideas were taken to an extreme: Everything in Boulogne was washed, steamed, sanitized, and disinfected. Operating rooms were immaculate; physicians and nurses were scrubbed; surgical instruments were spotless. Wounds were doused with increasing doses of ever-stronger antiseptics: phenol pastes, brilliant green dyes, solutions of bleaching powder, boric acid, perchloride of mercury, tincture of iodine. Every antiseptic was somewhat caustic. Some were terrifically painful. The process culminated in the Dakin-Carrel treatment and Sir Almroth's feeling of helplessness and failure. Listerism had reached its limits. Something else was required.

Five years after Almroth Wright's Boulogne laboratory was closed at the end of the war, Gerhard Domagk thought that he might be on the trail of the next great way to treat infection. He began staying all night in the lab at Greifswald, running experiment after experiment on his phagocytic cells and the substances they secreted. The university janitor in charge of the building where Domagk worked complained to his boss, Gross: Domagk was working too late, the janitor said; he needed to clean that space, and how could he with the young man working at all hours? Domagk was using too much electricity; experimental mice were getting loose. But Gross, impressed by his assistant's commitment and the quality of his experiments, protected him. In 1924, Domagk earned the title of professor, with Gross commending him for his "pronounced capability for fertile scientific questioning" and "remarkable independence." He began publishing papers. In 1925, when Gross moved to a better post at the University of Münster, he took Domagk with him. With renewed confidence about his future, Domagk finally married Gertrude.

The new position in a new part of Germany, to the south, in a more industrial area far from the sea, offered Domagk a bit more money but also some new challenges. He started thinking about everything in terms of tipping the balance in favor of the body: about white blood cells in general and his phagocytic cells in particular, how they could be used to control cancer and tuberculosis, how physicians might prevent anaphylactic shock and boost the immune response. He worked tirelessly on a half dozen fronts.

His position at Münster, however, was far from perfect. He was no longer a starving student, but neither was he a secure practicing physician. He was an academic pathologist, a doctor doing basic research within a university. He still worked under Gross, but Gross seemed to be getting less support at the University of Münster than he had at Greifswald, which meant fewer chances for advancement for his protégé. Domagk and his new wife were, he remembered, "poor as church mice," earning only enough to rent rooms in another family's house, with barely enough left at the end of the month to buy, if they were lucky, a bottle of wine. They determined to have children nonetheless. Their first son, Götz, was born in 1926, a sister, Hildegarde, three years later, then two more sons, Wolfgang and Jörg, in the early 1930s. And they were happy. "It was the right decision," Domagk wrote, "to work hard, to hunger, to marry and to go together through difficult times, because thus we never lost our good humor. Eventually it became better. Although there were stormy and rainy days, the sun never disappeared totally."

CHAPTER FIVE

POPULAR CULTURE fosters two myths about scientists. The first is the myth of the "mad scientist," a variation on the Prometheus theme in which the researcher seeks more power than the gods intend or uses nature's secrets to do evil rather than good. The result is a Dr. Frankenstein, a Dr. Octopus, or a Dr. No. The second is the myth of the selfless scientist: more intelligent and less ambitious than other humans, more independent and less prejudiced, more altruistic and less avaricious. These are the scientists who seek to bring the light of truth to mankind and ease its suffering: Dr. Schweitzer, Dr. Einstein, and, in literature, Dracula's nemesis Dr. Van Helsing.

The truth is that scientists come in all types, just like everyone else. They are people, not pop paradigms. They worry about how they are going to pay their bills, and they get envious of the researchers who got the credit they should have gotten. They compete for grants and complain when those grants are awarded to someone else. They focus on prestige and work for advancement and usually do what their bosses (or, less directly, granting agencies) say. Most scientists, as the great British molecular biologist J. D. Bernal noted back in the 1930s, live lives more like those of anxious middle managers than great visionaries.

Much of what real-world scientists do, for good or ill, comes down to money. Science even in Domagk's day (and now much more so) was a very expensive pursuit, so expensive that only a few organizations—notably government, industry, and a few large nonprofit groups—have the money to fully support it. As a result, most scientists are hired workers, civil servants, university professors, or corporate employees. Their search for truth is very often shaped and guided by their search for research funds.

That was why Domagk's future would be determined not only by his desire to stop disease but also by his own ambition, his family needs, and the plans of a small group of businessmen he had never met. He probably had heard of their leader, however, one of the preeminent figures in German business, a man the London *Times* would later eulogize as "the greatest industrialist the world has yet had." His name was Carl Duisberg.

Duisberg was a German version of Thomas Edison, Henry Ford, and John D. Rockefeller rolled into one. He had built an empire of science in Germany, leveraging the discoveries of dozens of chemists he employed into one of the most profitable businesses on earth. He knew how industrial science worked: He was himself a chemist. At least he had been long ago. Now, in the mid-1920s, in the twilight of his years, his fortunes made, his reputation assured, he often walked in his private park alone—still solidly built, with his shaved head and a bristling white mustache, still a commanding presence in his top hat and black overcoat—through acres of forest, fountains, classical statuary, around the pond in his full-scale Japanese garden, by the lacquered teahouse, over his streams, and across his lawns.

Through the trees he could see the smokestacks of his largest factory, Leverkusen, a wonder of his age. Leverkusen was Duisberg's masterpiece, the most modern, the most efficient, and one of the largest chemical plants in the world. It had taken a prominent German artist nine years to paint a huge mural of Leverkusen, with its hundreds of buildings spread across the better part of a square mile. Duisberg loved displaying the mural, which hung in his executive offices. With a wave of his hand, he could show visitors the vision that had inspired him. At the start, on the far left of the mural, ships could be seen steaming into his his docks on the Rhine, unloading raw materials to the railroad yard alongside them, delivering a steady flow of coal, packaging, and unprocessed chemicals to feed his factory. Moving to the right were the Bayer processing plants where the raw materials were pulverized and cooked into refined chemicals, which flowed farther right to a complex of buildings for testing and manufacturing into final products, which were transferred farther still to the right, to the packaging and storage facilities. Visitors could immediately see the sense of placing a second set of railroad tracks at the far right, ready to receive finished products and ship them to the world. In the trees at the very far right edge could be seen a neighborhood of executive houses, near enough to the factory so that Duisberg's managers could keep their fingers on the pulse of production, far enough away to provide his top men some peace and quiet. In the lower part of the painting was Duisberg's villa, near it the ornate Kasino, a combination hotel and private club for his managers and visitors, and next to the Kasino the mural's focal point—cleverly highlighted through the artist's use of light and shade—Duisberg's executive building, an intimidating neoclassical pile fronted with red sandstone, embellished with a huge Greek frieze, shining with gilt fittings. Leverkusen employed thirty-five hundred people, had its own bank and library, swimming pool and recreation hall, printing press, news operation, patent bureau, fire department, sports teams, and philharmonic orchestra. Duisberg had conceived it, funded it, and built it. There were no towns large enough and close enough to the plant to efficiently house his workers, so Duisberg built his own town next to the factory.

Leverkusen had been completed in 1912, just before the war. Now, a dozen years later, as Domagk was still just considering marriage, Duisberg was walking his grounds and thinking, preparing himself and his villa for a meeting he believed would be the capstone of his career, the greatest example of his renowned business acumen, and a milestone for Germany. Journalists and historians would later call it the "Council of the Gods."

All of this—the villa, the factories, the parks, the power—had started more than a century earlier in a pot of Scottish goo. The pot was in the home of a Scottish engineer and inveterate tinkerer named William Murdoch, a man who invented contraptions like steam-powered wooden bicycles, best known today for devising a process he called "distillation of coal," by which coal heated in a retort gave off quantities of a colorless gas that could be captured, piped to various locations, and burned in lamps. By 1792, Murdoch was using his "coal gas" to light his cottage and office. It was a distinct improvement over dripping candles and smelly kerosene lamps. It was the start of the Gaslight Era. The manufacture, sale, and distribution of coal gas quickly became a big business, something like electricity would prove to be a century later. But the distillation of coal did have its unpleasant side effects: There were occasional explosions, and pipes and retorts quickly became coated with a sticky black goo, "an offensive substance," one observer noted, "repugnant alike in appearance and in odor." It was called coal tar. Coal tar was a nuisance that had to be cleaned and disposed of, burned or buried. As demand for gas increased, so did the amount of tar.

It was not long before entrepreneurs found ways to make money off it. When heated and refined, the tar yielded subcomponents that could in turn be distilled and purified into scores of new chemicals. Through the early 1800s, coal-tar chemicals appeared in a steady stream: creosote (like the German Creosote that had caught Lister's attention), naphthalene, anthracene, aniline, carbolic acid, benzene. The list seemed endless. The study of these complex chemicals, all of which contained carbon, expanded into the field of organic chemistry, which quickly grew in importance along with the growing use of coal gas. To organic chemists Murdoch's goo was not garbage but a cornucopia of interesting molecules. Many coal-tar chemicals had immediate practical applications: One was a preservative for railroad ties, another a substitute for turpentine (which was made from tree sap); an inventor named Charles Macintosh figured out how to use coal-tar derivatives to make waterproof fabrics (the word "macintosh" became a synonym for a raincoat in Britain); other coal-tar compounds were used in everything from roofing materials to road binder. The chemists turned waste into gold; coal tar was an early demonstration of the power of industrial recycling.

William Henry Perkin, an eighteen-year-old English chemistry student, was another of those early researchers fascinated with coal tar. During one Easter vacation from his studies, he started looking for a way to use coal-tar chemicals to make artificial quinine—a very valuable drug, the only thing in the world that could treat malaria. It was 1856. Working in a crude laboratory in his home, treating coal-tar chemicals with other chemicals, heating, and cooling, Perkin failed utterly to find anything like quinine. But he stumbled onto something else along the way, a dark precipitate in the bottom of a flask that, when put into solution, yielded a lovely shade of purple. He found that he could dye silk with it. This was more than interesting. It was the first time cloth had been dyed a color not found in nature. Before Perkin only natural dyes were used: indigo from an Indian vegetable, carmine from a Mexican cactus, vermilion from insects, Roman purple from guano. Perkin's first dye, aniline purple, renamed "mauveine" for the marketplace, became the primary product in a little factory he started on the banks of Grand Union Canal in West London. His success was assured by a celebrity endorsement: Queen Victoria appeared at the Royal Exhibition of 1862 in a mauve silk gown, turning the new color into an international craze and its inventor into a rich man. After Perkin came the age of synthetic colors, a rainbow of artificial hues more colorful, more colorfast, and more profitable than anyone dreamed possible. Perkin himself went on to discover Perkin's Green, Britannia Violet, and alizarin red, among other coal-tar dyes. Soon it was said that the water in Grand Union Canal changed color every week, depending on what shade Perkin was making.

By then, though, the center of the synthetic dye industry had moved. While Britain and France battled each other over patents for the new dyes, Germany—a nation with both the motivation (few abundant natural resources other than coal, water, wood, and air) and the expertise (the best chemists in the world)—capitalized on Perkin's discovery and took over the synthetic dye industry. British Manchester Brown and French Magenta (named after a military victory of Napoleon III) were outsold by German synthetic indigo and Congo Red. The French and the English still regarded chemistry generally as the pursuit of leisured gentlemen. The Germans saw it as a way to international power. Germany was a new nation, unified fully only in the 1870s, and was too late on the scene to build the type of colonial empire Britain had. Germany had to compete in other ways, and science was the key. The Germans created the world's best science departments in their schools, encouraged cooperation between university researchers and private industry, and threw government money into the pursuit of valuable new insights and processes. They were innovative in their science—finding new ways to chemically alter coal-tar chemicals to create brighter and better colors—efficient in their production techniques, and aggressive in their marketing. A huge coal-tar industry sprang up in Germany in the last half of the nineteenth century. The new German coal-tar colors were immensely popular; the firms that made them, the high-tech companies of their day, quickly grew into giants.

ONE OF THE most important was Friedrich Bayer & Co., started in the 1860s by the son of a silk weaver. Bayer's first coal-tar dyes were products of kitchen chemistry brewed up in earthenware pots on a stove; he sometimes threw in ingredients like egg whites in an attempt to improve his recipes. His small firm discovered how to make a few dyes that proved profitable in the rapidly growing market and hired an aggressive business manager named Rumpff—a man with the good sense to marry Bayer's daughter—who believed in finding many more. Like any able business manager, Rumpff wanted above all things to make money and was convinced that in order to do so the firm needed real scientists, chemists who could efficiently discover new colors. In 1882, he persuaded the Bayer board to hire three young chemists at very modest salaries. One of them was Carl Duisberg, twenty-one years old, strong and handsome, with a luxurious brown mustache and boundless self-confidence. He was not considered brilliant in the laboratory, but what he lacked in chemical talent he made up for in ambition. At Bayer he worked hard, got lucky—discovering one new dye by accident, another, he said, through a dream—and he, too, had good sense. He married Rumpff's niece. Duisberg climbed quickly: laboratory manager at age twenty-five; board of management before he was forty. He designed the Leverkusen plant, and in 1912, the year it was completed, just past his fiftieth birthday, he assumed full control of the company.

During that climb he moved away from chemistry and toward his real talent: industrial management. Duisberg was a visionary and a leader, a man who loved his company, treated his workers like members of his extended family, and yet was open to new ideas from other cultures. He loved to travel. In 1903, while on a trip to the United States to review the site for a new Bayer factory, Duisberg trekked from Canada to New Orleans sampling American food and studying American business practices. He returned to Germany with unbounded admiration for giant trusts like Standard Oil, combines that offered, he noted approvingly, "the merger of similar operations under a single management and supervision and with a single sales organization to remove competition of all sorts of products—particularly those subject to price pressure—a way to increase profits without a substantial rise in sales prices." What worked for the Rockefellers, Duisberg believed, could work for German dyemaking firms, which by then had diversified into far more than dyes, making more money off the discovery that coal-tar substances could be chemically altered to make many other products, from photographic chemicals to pharmaceuticals.

Once home from the United States, Duisberg set up meetings between himself and representatives of Germany's two largest dye and chemical firms (Hoechst and Badische Anilin- und Soda-Fabrik, better known as BASF; Duisberg's company, Bayer, was number three). Their executives listened politely to the ambitious young executive, then asked him to write down his thoughts. The result was a fifty-eight-page memo describing the future of Germany's chemical industries. German chemical research was by all measures the best in the world, he wrote. German chemical industries were the most innovative and wide-ranging in their activities. German sales forces dominated global markets. But despite all these advantages, these same German firms were undercutting one another's prices, duplicating one another's research, poaching one another's patents, and tearing one another to pieces. The smallest firms were dying. The largest were not realizing all possible profits. Opportunities and riches were slipping away. The solution, he wrote, was simple: Bind the competing companies together. Join the competitors into a single family. A merged German chemical giant, moving in a single direction, freed of the crippling costs of duplicate research and marketplace competition, able to coordinate production and lower management costs, would reap staggering profits.

Duisberg's memo, one executive exclaimed, was "a masterpiece born of genius." But Hoechst, proud of its independence and confident in its ability to discover new products, decided not to join. Duisberg focused on BASF. The two firms sized each other up, growing comfortable enough to exchange financial data. Just as they were nearing a merger, Duisberg read in the newspaper that Hoechst was forming its own cooperative arrangement with another major chemical player. In response Bayer and BASF quickly reached an agreement. In 1905 they were joined by a third firm, Agfa, creating the "Triple Confederation" of Bayer, BASF, and Agfa. Hoechst countered with a third firm of its own, creating a "Tripartite Association" in 1907. The two triplets joined forces temporarily during World War I, creating something more of a common-interest group than a merged supercorporation. By the time the war was over, the firms were beginning to appreciate the value of cooperation. Things were moving in Duisberg's direction.

Now, in 1924, the time was right to complete the process. The German dye companies had survived crisis after crisis during the war: last year's major export markets turning into this year's enemy states; half the employees drafted; the "Turnip Winter" of 1916–17, when the British blockade began to have its effect and Bayer's starving workers went on strike. British troops had occupied Duisberg's villa, moving the magnate and his family into two rooms and some cellar space. In the days after the war, the Left began organizing Bayer's labor force. Public finances in Germany were disrupted, universities were in disarray, and scientific research was in decline. At the same time, worldwide competition was heating up. Other nations, denied German chemicals during the war, had boosted their own research and expanded their own chemical industries, finding ways to make their own dyes, drugs, and fine chemicals. Export markets were drying up. After the war the victors took, among the spoils, many of the assets, patents, and even trademarks from Bayer and the others. An American patent-medicine maker best known for laxatives, dandruff remedies, and impotence cures purchased all of Bayer's U.S. holdings, including the vaunted and immensely valuable Bayer Aspirin brand (Bayer Aspirin sold today in the United States is made not by Bayer but by Sterling Products). German science, the best in the world before the war, was knocked off its pedestal. Then came what looked like the final blow: inflation. Before the war it took about 4 German marks to buy a dollar. At the beginning of 1920, it was 49 marks. Two years later it was 188 marks and spiraling out of control. "Now that the South Sea islanders' cowrie shell is worth more than the paper mark, we are no longer at the edge of a precipice," Duisberg told his workers. "We have long since fallen off." But he had not seen the worst of it. In 1923, the mark traded at 49,000 to the dollar in January, 126 million in September, 72.5 billion at the end of October, 4.2 *trillion* to the dollar on November 20 of that year. A pound of bread cost 800 million marks. A pound of butter 20 billion. Bayer's 1923 balance sheet ended with a number twenty digits long.

But through it all, the dye and chemical makers not only survived but somehow managed to prosper. They profited even from the ruinous inflation, which allowed them to pay off old debts with devalued currency and reap windfall profits in foreign currency. Inflation at home meant they could charge more competitive prices abroad, which allowed them to fight off the worst of the market incursions by foreign chemical companies. Bayer's factories, undamaged during the war, ran at full capacity. Adroit managers like Duisberg kept their workers happy enough, even finding food for them when times were worst, giving them a reason to turn their back on the Communists. Sales were healthy for the most part, profits piled up, and sufficient reserves accumulated to let the companies ride out labor unrest and even begin to explore new products—artificial fertilizers, artificial rubber, artificial gasoline, new drugs—to offset the lost patents and a declining demand for synthetic dyes. By 1924, adjusted for inflation, the aggregate worth of German chemical companies was triple what it was before the war. Now, thanks to the financial and political unrest, changes were being made in Germany's corporate and tax laws favoring the formation of larger organizations. The time was right to discuss a total merger.

Duisberg called his Council of the Gods in 1924 to talk about next steps. The two major players were Duisberg and Carl Bosch, head of BASF, a dye company that had grown rich by industrializing a process to pull nitrogen out of the air and put it into fertilizers and explosives. The two men could not have been more different. Duisberg was now seen as a grand old man of the prewar period, a bit old-fashioned, a bit pompous. In his villa he kept a "Wall of Fame" of his framed awards and diplomas. He liked to list all of his degrees, honorary and otherwise, on his business cards. (One story bandied about the industry concerned the time Duisberg had traveled to meet with a high government official and had been kept waiting outside his office. When the meeting finally started, Duisberg complained that he was not accustomed to cooling his heels in anterooms. "Do forgive me," the official replied. "It took me so long to read your calling card.") Bosch, on the other hand, was one of the new breed of postwar technocrats, as quiet, efficient, and melancholy as Duisberg was forceful and avuncular. Duisberg was an indifferent researcher; Bosch was a Nobel-caliber scientist (he would win one in 1931 for his work on nitrogen fixation). Duisberg and Bosch did not much like one another. As soon as the council started, the two men began jockeying for position. Two factions formed around them, the smaller favoring Duisberg, who wanted to devise a detailed plan in advance for every aspect of the proposed merger, the larger group gathering around Bosch, who championed a looser, more improvisatory structure. At dinner the discussion became heated. Afterward Bosch's supporters retreated to the villa's bar while Duisberg's gathered in the billiards room. A mediator rushed from one group to the other, presenting proposals and counterproposals.

Bosch prevailed. At the end of two days of discussions, it was decided that Germany's foremost chemical companies would fuse into a single giant cartel, formed along the lines Bosch wanted, with Bosch at its head. Duisberg was named chairman of the board. They christened it the Interessengemeinschaft Farbenindustrie Aktiengesellschaft (Interest Community of the Dye Industry, Inc.). It soon would become internationally infamous under a shortened name: IG Farben.

Nothing like it had been seen before. IG Farben was on the day of its birth the largest corporation in Germany, the largest business in Europe, the largest and most powerful chemical company in the world. It was also the world's third-largest business of any sort measured by numbers of employees, bested in the mid-1920s only by U.S. Steel and General Motors. As Duisberg hoped, the IG Farben structure led to coordinated research and rationalized production; member firms began complementing one another's efforts rather than duplicating them; resources were freed to invest in the next big products to generate the next big profits. At Bayer those products would include a number of new medicines.

There was good reason for Bayer to focus on pharmaceuticals. With the dye market shrinking, drugs beckoned as a growth area richer than almost any other. The formation of IG Farben and the attendant specialization of product lines freed Bayer to put more emphasis on developing synthetic drugs, an area in which it had a strong track record and which had excellent growth prospects. Duisberg needed new products, new patents, to replace the losses of the war. The new effort would require new talent—including that of an idealistic young research physician named Gerhard Domagk.

CHAPTER SIX

By 1927, Domagk realized that the University of Münster was not working out as he had hoped. True, he had his own laboratory for the first time, was making a little more money (although still barely enough to support himself, Gertrude, and their new son), and was productive enough to begin publishing papers on his research into phagocytic cells, publications that began to earn him a reputation in his field. But he wanted more, especially after the baby came.

Münster was looking like a dead end. Gross was a good boss, but not much money was being funneled to his institute of pathology at the university, and the lack of funds limited the types of research Domagk could do. Nor did he see a clear route to future advancement. Gross was young, and he was not going anywhere soon. With Gross in place, Domagk could not rise. The Domagk family's unofficial motto—"Work, work, and go hungry"—had not changed, but even so, two years after moving to Münster, Domagk was growing restless.

He was becoming known for his ideas on weakening bacteria as a way to improve the odds of fighting off disease, but the initial burst of enthusiasm he had felt with the move was giving way to more mundane worries. "Should I not have enrolled for practical medicine in order to gain proper money?" he wrote in his journal. "Why did I have to go into theoretical research, particularly into experimental pathology in which there is no money? No one in the circle of university pathologists seems to be interested in me at all, and no one can possibly sustain a family just from doing autopsies and teaching." Gross, decent and supportive as he was, could not offer Domagk much hope; he was not in a strong enough position within the university to offer his protégé either a significant raise or a path to an academic chair. Both of them were passionate about pathology, but now Domagk worried that he had made a serious mistake trying to carve a career for himself in an academic backwater. "Pathology is only a side branch," he wrote. "Everyone says it is important, but it does not lead you anywhere and it does not pay off."

Then he received an offer that would give him almost everything he wanted.

It was from a man named Heinrich Hörlein. He let Domagk know that he had been reading his publications on the immune system with interest and had been impressed with Domagk's ideas about the interactions between the body and bacteria. Hörlein was the head of the pharmaceutical research program at Bayer, headquartered in a factory not far from Münster. Perhaps Domagk had heard of the man who ran Bayer, Carl Duisberg? His firm was expanding its new drug research under Hörlein's direction, and part of the strategy involved a new program to test chemicals for medical effects in animals. Hörlein had been given the go-ahead to construct a new research building, including a state-of-the-art laboratory for pathology. Whoever directed that laboratory would have control of what might be the best new-drug research facility in the world, would be given a number of laboratory assistants, would have access to a large stock of test animals, and, of course, would have the chance to cooperate with the great minds within one of the largest industrial/chemical companies in the world. They were looking for someone young and talented to start this new program, someone with experience in both medicine and animal testing. The person they hired would have a title something like "Director of Experimental Pathology." There would be a significant salary. Hörlein did not have to say, because Domagk understood this without being told, that working at Bayer also meant access to what seemed, in comparison to the support at his university, enormous sums of money to support his experiments. It would be a small program to start, but Bayer was quite interested in this area of research; it promised great things. Would Domagk be interested?

Domagk immediately responded that he was very interested indeed. His reaction might have seemed odd to other university researchers in other nations. Doing science for a corporation was disdained by most academic scientists, who believed that only in a university setting or perhaps a government laboratory could a scientist follow the trail of pure knowledge, unsullied by commercial concerns. In Germany, however, the situation was different. German science had become the best in the world because German schools were among the best in the world, and German schools tended to have productive relationships with German industry. The nation's rise to become the world's greatest center for research had started with fundamental education reform in the early nineteenth century, a time when German universities rooted out ancient theory-based approaches, substituting a central faith in empirical research and experimental discovery. Laboratories became an integral part of universities. The guiding principle was simple and powerful: Before you could know, you must prove. Science students in this new German system graduated well versed in the latest laboratory techniques and discoveries and well suited to working in the growing dye and chemical industries. Until World War I broke the German monopoly on chemistry, no matter where you were in the world, you could not consider yourself a chemist (or much of a physicist, for that matter) until you first spent time in Germany studying with a master. Scientists from around the world flocked to Germany and came home to remake their own colleges. Johns Hopkins, founded in 1876, was the first German-model school in the United States, the first "research university." Hopkins introduced many German-style innovations into American education: undergraduate "majors" instead of a generic liberal arts degree; small seminars with their give-and-take with a professor in addition to lectures; an emphasis on original faculty research, especially in the sciences; "doctoral" degrees awarded to students once they had shown their own ability for independent and innovative inquiry. Soon virtually every major university in the United States was doing what Hopkins did, instituting policies that had been in place in Germany for a generation.

Good schools were part of the fabric of German life. Even before

German unification, separate states had vied with each other to create the best *technische Hochschulen* to train workers. They competed with each other for the best minds, luring scientists to their universities the way former princes had lured composers and artists to their palaces. It was not unknown for chemists to be elevated to the German nobility.

Because the Germans believed in the value of linking their university programs to the needs of industry, the nation's academic scientists were not considered automatically superior to industrial scientists—a prejudice all too common in the United Kingdom and the United States. On the contrary, German industrial positions often offered top scientists better pay, facilities, and equipment. Industrial funds supported academic research; academic discoveries fueled industrial growth. The best men could be found in either place. Some could be found in both.

A prime example was a genius of the previous generation who had inspired both Domagk and Hörlein (as well as almost every other medical scientist in Germany): Paul Ehrlich. A contemporary of Duisberg's, Ehrlich was interested less in how to find new dyes than in how to use them in new ways. Even as a young man in the 1870s, Ehrlich had been fascinated with dyes, spending so much time playing with them as a student that his classmates called him "the Man with Blue, Yellow, Red, and Green Fingers." He saw dyes as a way to join the three great loves of his life: the study of bacteria, the study of medicine, and the study of chemistry. They would also become his route to a Nobel Prize. A German Jew born to an innkeeper in Silesia (now part of Poland), Ehrlich decided on medicine for his profession but became interested along the way in the exciting news about bacteria causing human disease, in seeing them under the microscope, and in developing better ways to see them. The microscope was every bacteriologist's central instrument, but it had a terrible drawback: Much of the microscopic world could not be seen, because it was more or less transparent. Under a microscope water was transparent, serum was transparent, white blood cells looked like clear blobs, bacteria were ghosts. Seeing detail in tissue slices was almost impossible—everything blended into everything else.

The solution was dye. Properly stained with some of the new synthetic dyes chemists were making from coal tar, human cells and bacteria popped out of the background, showing cell walls, nuclei, granules, tiny organs, inner structure. Certain dyes stained certain structures preferentially; combinations of dyes could be used to differentiate between different cells or structures within cells. While studying medicine, Ehrlich spent more time using dyes to stain slices of tissue than he did memorizing the parts of the body. He invented new stains specifically for bacteria that gained wide use, and he developed improved ways of using others. Ehrlich earned his doctorate with a dissertation on the practice of using dyes to stain animal tissues.

He was especially interested in the fact that different dyes stained different kinds of tissues. They were somehow matched. You could stain bacteria one color, for instance, and the human tissue around it another. This was something that cloth dyers already knew—some dyes stuck better to wool, for instance, than to cotton—but scientists had not explored how or why. Ehrlich made it his life's work. He figured that for a dye to be effective, something specific in its chemical structure had to latch on to something specific in the cells. Ehrlich, part physician, part bacteriologist, part chemist, wondered if it might not be possible to work with that relationship in a medical way, to find a dye, say, that would bind specifically and strongly to a disease-causing organism, to a bacterium, and not to human tissue. If one could be found, might it not be possible to attach a poison to the dye and kill bacteria with great specificity even inside the body? The poisonous dyes could work like magic, distinguishing pathogen from patient, picking out and killing the bacteria only, making it possible to safely disinfect a human from the inside. Regular antiseptics, the kinds Lister used, were too harsh to use inside the body. But Ehrlich's dyes might. He imagined them working, could see no reason that they might not, even made up a name for his imaginary new medicines: *Zauberkugeln,* "magic balls."

He chased the idea through years of research with his mentor, Robert Koch, making the stains Koch used to find the bacteria that caused tuberculosis (Ehrlich in the process coming down with the disease himself; he had to spend two years recovering in Egypt). He and

Emil von Behring, another of the group of young researchers gathered around Koch, made a special study of diphtheria. Together they perfected the way to stop the disease by injecting an antitoxin, a serum made in horses. Ehrlich knew a great deal about the immune system, and he was learning more all the time about disease-causing bacteria from Koch. He became adept at finding and purifying cultures of these organisms, studying them in animals, and testing the ability of drugs to kill them in test tubes. Always in the back of his mind was the dream of *Zauberkugeln*.

Ehrlich searched for his magic medicines obsessively, on his own, lost in thought. He ate little, smoked twenty-five cigars a day—a box often under his arm, ashes flecking his unkempt clothes—and scribbled notes on any scrap of paper he could find. If there was no paper handy, he wrote his ideas on laboratory walls, shirt cuffs, and tablecloths. He worked for three years without a salary. His laboratory was a mess, a jumble of test tubes, papers, bottles, and journals stinking of cigar smoke. Once a housekeeper made the mistake of trying to tidy up his office; when he saw what she had done, he told her that he had hidden poisons among his papers to which there were no antidotes. As he planned, she stopped touching his things. People thought he was crazy. Soon he had a new nickname: "Dr. Phantasmus."

If he had been nothing but a dreamer, his work would never have mattered. But Ehrlich's ideas were firmly grounded, his observations precise, his techniques innovative. He was a demon for analysis and measurement. He designed his theories as carefully as an architect blueprints a house. "His exactness in calculation was absolute," one historian wrote, "and, to some of his colleagues, appalling."

He earned fame for his meticulous studies of diphtheria antitoxin and the immune system. And eventually the rewards started coming. In 1891, he was made a professor at Berlin University, in 1896 was put in charge of a serum research center, in 1899 was made director of the Institute for Experimental Therapy in Frankfurt-on-Main, and a few years later was given charge of the Georg Speyer House, a private research institute. In 1908, he won the Nobel Prize for his work on immunity and serum therapy. His greatest triumph was yet to come.

He was still searching for his *Zauberkugeln*. He cared not at all that every experiment with internal antisepsis had ended in disaster, that no one had ever seen a chemical injected into the body magically pick out and kill bacteria without harming the tissues around it.

Ehrlich believed in his ideas because he had seen something most physicians had not. He had seen blue nerves. Using the dye methylene blue, he had treated a nerve-rich tissue sample and seen the delicate neural networks emerge out of the background shaded a beautiful indigo, a meshwork of fine filaments in detail never before seen. If methylene blue could pick out and highlight nerve cells within the crowded curiosity shop of the body, he was convinced that some dye—perhaps some variation of the same dye—could recognize and attach to disease-causing organisms as well. Besides, there was another reason to believe: The body already was making its own sort of *Zauberkugeln,* magic molecules called antibodies, proteins that somehow were able to recognize and aid in the removal of specific bodily invaders. "The living cell," he wrote, "is nothing but a small chemical plant." If those little factories could make *Zauberkugeln,* science should be able to as well. Ehrlich experimented with methylene blue as a medicine, figuring that its preference for nerve cells might give it some effect as a painkiller. He also tested it against cancer and malaria. The dye had little effect in animal tests, but it did have some. Enough to give him hope. He kept searching.

Then he devised a way to improve his odds by ratcheting up the speed and scope of the search. He hired chemists to assist him, started them off with a promising molecule (like methylene blue or another dye), then had them make chemical variation after chemical variation, keeping the starting molecule's core structure but tinkering with it, varying first this part of the substance, then that, performing molecular makeovers using many of the same techniques chemists had been using for years to find new shades of dye. Ehrlich then tested each new variation against diseases in animals and bacteria in test tubes.

Under his close supervision—stacks of handwritten notes, clouds of cigar ash, flashes of temper—his team produced and tested hundreds of new chemicals. The most promising were those containing arsenic—arsenicals, they were called. Ehrlich heard of an arsenical named atoxyl that when tested in mice had cured cases of African sleeping sickness (a disease so deadly, so widespread south of the Sahara that it scared most white colonizers away from the interior). Unfortunately, atoxyl was too toxic to use in humans. So Ehrlich set his chemists to work synthesizing variations, one after another. He tested a hundred versions of altered atoxyl on mice. Then a hundred more. And two hundred more. Number 418 looked so promising that Ehrlich announced to the world that he had found a cure for sleeping sickness. But he spoke too soon. Number 418, also, proved too toxic for general use. He and his chemists resumed the search.

Ehrlich said his method consisted basically of "examining and sweating"—and his coworkers joked that Ehrlich examined while they sweated. There was another motto attributed to Ehrlich's lab, the list of "Four Gs" needed for success: *Geduld, Geschick, Glück, Geld*—patience, skill, luck, and money.

Patience and skill they had. Money was coming from both the German government, intent on conquering sleeping sickness as a route to colonial expansion in Africa, and the Hoechst pharmaceutical company, which had a close business relationship with Ehrlich's lab. What they needed was luck. And that, too, came their way, thanks to one of Ehrlich's rare mistakes. A few years earlier, the germ that caused syphilis had been isolated—a strange, curled, squirming microorganism that seemed to be part parasite and part bacterium—and Ehrlich thought (wrongly, as it turned out) that the syphilis germ was closely related to the parasite that caused sleeping sickness. Syphilis was a bad disease, but the germs were less dangerous to work with in the lab than were the organisms that caused sleeping sickness. Ehrlich found that he could infect lab animals reliably with syphilis, which gave him a way to test his chemicals against the disease. He had a hardworking assistant from Japan named Hata who had some experience doing work on syphilis in lab animals, so Ehrlich assigned him the task of going back and retesting every chemical they had already tried on sleeping sickness, this time on syphilis. By now the series was into the 900s. In 1909, Hata found, to Ehrlich's surprise, that number 606—another failure with sleeping sickness—could reliably cure syphilis in rabbits. They announced their finding to the world in 1910.

The news created a sensation. Syphilis was a legendary and much-feared disease, far more important to Europeans than was sleeping sickness. Syphilis, it was said, had afflicted everyone from Henry VIII to Oscar Wilde. It was a horrible disease that discomforted, then disfigured, then maddened, then killed. The rueful phrase "a night of Venus and a lifetime of Mercury" summarized both the disease's cause and its sole treatment: a lengthy regimen of dangerous doses of mercury, a toxic heavy metal that, while it did not cure the disease, at least slowed its progress. For centuries mercury had been the only available treatment.

Ehrlich's 606 was a distinct improvement. His staff made it in quantities large enough to give to physicians for testing in humans; Ehrlich himself oversaw the distribution. The results were positive. Word spread that there was a cure for syphilis, demand soared, and Hoechst quickly began selling it under the name Salvarsan.

Salvarsan was a great advance in medicine, but it was not a true *Zauberkugel*. Ehrlich had set the criteria himself: The perfect internal antiseptic would harm only the invading microorganism, leaving the human patient untouched. Salvarsan was so toxic that it could be administered no more than once a week. It had to be given intravenously, because a muscular injection would damage the flesh at the site of the shot; the shots were painful; it caused rashes and liver damage; long-term treatment could lead to jaundice; sometimes patients died before their disease was cured. The drug's side effects were so appalling that many patients refused to finish the course of treatment. Ehrlich quickly began looking for a less toxic chemical variant and found one: Neosalvarsan was released by Hoechst in 1912. Journalists writing about Ehrlich's medicines later came up with a better name for what he had found. They likened the way his drugs worked to policemen chasing a criminal into a crowd, armed with weapons that, when fired, unerringly sought out and killed only the criminal, sparing the innocent. The new nickname for Ehrlich's medicine was "magic bullet."

Salvarsan and Neosalvarsan were imperfect, but they did prove one thing: Chemicals fashioned in the laboratory—entirely synthetic substances—could cure disease. The Hoechst firm made a fortune off of the discovery; the new antisyphilitics helped it grow into the most prosperous drugmaker in Germany.

Ehrlich himself won a Nobel Prize, was lauded around the world, was featured in a bestselling book, *The Microbe Hunters,* in the 1920s, and was the subject of a major feature film, *Dr. Ehrlich's Magic Bullet,* released in 1940. But that was long after he had died, a deeply unhappy man. Despite the euphoria of 1910, despite his optimistic announcements of coming miracles, Ehrlich never found another Salvarsan. All the new drugs he tested between 1910 and his death in 1915 were failures. Some say Ehrlich's death was hastened by critics, some of them physicians, who called him a "quack and a murderer" because of Salvarsan's severe side effects. Ehrlich began drinking heavily and died of a stroke when he was sixty-one. He was buried in the Jewish cemetery in Frankfurt. His grave was later desecrated by the Nazis.

Nonetheless, Ehrlich's achievements and the power of his central idea—chemical magic bullets could clean invading organisms out of the body—continued to guide medical research through the 1920s. If there was a Salvarsan for one disease, why not more? Other German university researchers and chemical companies—including Bayer—began emulating Ehrlich's approach, searching for the next magic bullet. After Salvarsan there was a burst of optimism, a sense that chemical cures would quickly be found for each disease.

But the optimism soon soured. Drug after drug was announced with great fanfare. Test after test was made. And almost every one would be found wanting, either too toxic to use or too weak to cure.

Optochin, announced by Ehrlich as a cure for pneumonia in 1911, looked as if it was going to be the next in a parade of medical wonders. It certainly looked promising in mice, and early human tests were encouraging. Instead Optochin convinced at least some scientists that synthetic chemicals would never cure disease.

Sir Almroth Wright's experience with Optochin robbed him of his faith in drugs. It happened in the diamond mines of South Africa. The miners there around 1910 began suffering from an epidemic of pneumonia, a perfect test for Ehrlich's new drug. Sir Almroth arrived in 1911, armed with Optochin and accompanied by his right-hand man, Leonard Colebrook. The diamond miners offered a number of advantages when it came to human testing: They were carefully supervised; they could be ordered to take their medicine; and once they had taken their medicine, they could be ordered to submit to exams and blood sampling.

At first it looked as if the new drug was working. After injecting it, Sir Almroth's group could detect it circulating in the blood of the miners, and it appeared to be having an effect on the bacteria causing pneumonia, although the results were mixed and complete "cures" were few. Then the problems started. Miners started arriving for their checkups complaining of blurry vision. In some cases it became severe. Then some of the patients went blind. Wright and Colebrook, to their horror, realized that the drug they were injecting into the miners was damaging their optic nerves. The blinding was irreversible. When he saw what was happening, Sir Almroth cut the tests short and hurried back to England. Although he knew and respected Ehrlich—as a young man, he had studied under the great physician/researcher—Optochin and the blinded South African miners made up his mind. Sir Almroth Wright spent the rest of his life explaining to whoever would listen the impossibility of effective drug therapy. He was convinced that Ehrlich and the Germans were wrong. He was steadfast in his belief that the only sure route to health was through the immune system. He was certain that chemicals would never cure disease.

Later his critics gave him a new nickname: "Sir Almost Wright."

CHAPTER SEVEN

BY THE MID-1920S, most researchers who had sought to follow in Ehrlich's footsteps had become pessimists. Only a few laboratories kept up a significant search for magic bullets. One of the largest of these, a fiefdom within Duisberg's enormous Bayer empire, was run by Heinrich Hörlein.

Negotiations between Gerhard Domagk and the Bayer Corporation went smoothly. Hörlein very much wanted the young pathologist for his expanding research program to search for new drugs, and the young pathologist very much wanted a higher salary and expanded lab. Domagk and Hörlein agreed on money and title as well as a general line of inquiry. Hörlein wanted magic-bullet drugs that could stop common bacterial infections inside the body the way Ehrlich's Salvarsan had stopped that single micro-oddity, the syphilis germ. Hörlein believed in the dream of internal antisepsis. Domagk also got an okay to test new chemicals against cancer, an area in which his approach of finding not necessarily a killer drug but a wounding drug—something that would damage the invading organisms or cancer cells just enough to tip the balance in the body's favor—might work. A two-year renewable contract was signed, long enough to give both parties time to see if they could create a compatible and productive team but short enough to ensure that Bayer would not be committed to the young physician if he failed to perform up to expectations. Domagk briefly tried to leverage the offer into a more attractive deal for himself with his current employer, the University of Münster, but the school did not have the resources to compete. Münster did not want to completely lose a young man beginning to look like a rising star, however, and kept Domagk on the books as a lecturer, giving him an open door to return if he wanted, a safety net if Bayer did not work out. Then Gerhard Domagk, age thirty-one, packed up his wife and infant child and moved south to the Wupper River Valley in Germany's industrial Rhineland, site of the Bayer plant at Elberfeld and home to a new phase in his life.

HEINRICH HÖRLEIN, with his round glasses, his three-piece suits, his mousy mustache, and his paunch, looked more like a mild middle manager than a revolutionary. He was in many respects the epitome of a company man. But he was also, in his own way, intent on igniting a new era in medical research.

Hörlein was a lifer at Bayer who had worked his way up the ladder for twenty years. He was hired as a chemist, did a bit of dye research, but made his name by discovering Luminal, a drug for epileptics that also became a popular sleeping medication. Luminal was a big seller for Bayer. Hörlein's talents were recognized, he was brought into management, and now, in 1927, he was a top executive, member of the executive board, head of the technical committee, and the man the company put in charge of chemical, bacteriological, and pharmaceutical research. A longtime admirer of Paul Ehrlich, Hörlein had been deeply impressed by the Salvarsan triumph both scientifically and financially. He attributed the subsequent failure to find more drugs to errors not in approach but in scale. Hörlein understood the challenge: While researchers knew quite a bit about the various families of bacteria, they still knew very little about how bacteria worked at the chemical level, how their metabolisms operated, or how they reproduced. Without this knowledge it was impossible to "design" drugs to fight them. Instead groups like Hörlein's had to search widely, screen hundreds, perhaps thousands, of chemicals in hopes of finding one that worked, expect constant failure, and keep searching. They were groping in the dark, but to this hit-and-miss approach Hörlein added size, power, and a dose of educated guessing. They would start with a promising compound, anything that looked as though it might have some antibacterial activity and yet not be too toxic to give to a living animal, then start playing with it, chemically altering it hundreds of times, testing each new variant against a variety of bacterial infections in animals to see if anything worked. It was a bit like drilling for oil: You figured something must be down there, you looked for the most promising geology, but to find out for sure you had to sink a well. Most wells would be dry, but the payoff from a single gusher could make you rich. Once you found one, once you tapped in to an oil field, you could drill more wells nearby with a higher chance of success. Drug research in the 1920s was like that. The vast majority of the substances tested did the expected—that is, nothing—but there was still the chance to find another Salvarsan and make a fortune. Ehrlich, Hörlein believed, had pioneered the right method. Only one thing was needed: another jump in scale, a ramping-up of the discovery process to full industrial levels. If fifteen years of failure after Salvarsan showed anything, it showed that single investigators or small teams working on one or two chemicals a week were not going to find anything. Successful drugs were apparently rarer than that. How rare, no one really knew. What diseases they would conquer, no one knew. How long it would take, how much money, no one knew. Hörlein intended to find out.

Domagk was hired into a system for prospecting new drugs that had been evolving and growing at Bayer for almost twenty years. The overall leader was Carl Duisberg, of course, Bayer's top executive, an astute businessman who recognized the profits possible from new drugs and who funded efforts to find them at astonishing levels, especially after the war. There was his right-hand man for drug research, Hörlein, who designed the system and hired the workers. Then there were chemists, physician/researchers like Domagk, laboratory assistants, animal-care workers, secretaries, and so on.

Drug development was a team effort. But it is safe to say that Domagk worked in an environment that existed most directly because of the ideas and management style of Heinrich Hörlein. Hörlein had come to Bayer in 1909 for the same reason Duisberg had: to find improved dyes. One of his first jobs as a recently graduated young organic chemist was to look for ways to make dyes more colorfast, less liable to fade in the sun or with repeated washing. Among his earliest discoveries was that adding newly found side chain with sulfur in it—para-amino-benzene-sulfonamide was its scientific name—to certain dyes made them appreciably more colorfast. He and his co-chemists made a few of them, oranges and yellows. But Hörlein was interested in more than colors. He started making suggestions for improving laboratory organization, he showed some talent as a manager, he was a team player, and he was quickly promoted. Two years after he was hired, at age twenty-nine, he was made head of Bayer's pharmaceutical department and began hiring chemists himself. He was soon put in charge of Bayer's research laboratory complex at Elberfeld, the company's think tank and testing site. Elberfeld was old and outdated (once Bayer's main dye factory, it had been superseded by the modern new factory at Leverkusen). But it was still huge, still fully functional, still full of possibilities. Elberfeld became Hörlein's personal domain. He was determined to make great things happen there.

He had Duisberg's trust. After Salvarsan showed how much money could be generated from synthetic drugs—and with Bayer's own mysterious pain and fever reducer Aspirin generating huge profits worldwide (no one understood why it worked; it just did)—Duisberg decided to invest in a big way in new drug research. That meant money for Hörlein, who sketched out a grand plan for finding new medicines, secured funds for buildings and equipment, and began searching for talent. Hörlein's first coup was to hire a superstar of drug research, one of Ehrlich's former top assistants, Wilhelm Roehl.

A young physician with an intellectual's high forehead but a gangster's heavy features—picture Edward G. Robinson in a lab coat—Roehl arrived at Bayer at the end of 1909 with all the talent, ambition, and discipline needed to be the next Ehrlich. He brought with him many of the systems and techniques he had learned in Ehrlich's laboratory, especially Ehrlich's methods for testing new chemicals on animals. Bayer, then still primarily a dye company, was at the time ill-prepared for this new line of research. Roehl was exiled to a whitewashed plumber's shack outside the plant walls (outside, some said, because of a lack of space; others said it was fear of the disease germs he was brewing). But it was enough. He gutted it, fitted it, and was given leave to staff it as he saw fit.

Of course, he was not given a blank check. Ehrlich had done his work with minimal equipment, and Roehl was expected to do the same. So he cleaned and arranged, turned the shack into a workable laboratory, laid claim to the one decent microscope he could find at Bayer, and hired assistants whom he trained so thoroughly that they could run experiments on their own. Then he started searching for new medicines.

Out of consideration for his mentor Ehrlich, who was then looking for the next Salvarsan, Roehl steered clear of arsenicals. He concentrated instead on a family of dyes that Ehrlich had shown to have some effect against the organisms that caused sleeping sickness, although not enough to make a profitable medicine. They were called azo dyes: afridol violet and sulphur red acid among them. Ehrlich was using mice to study sleeping sickness, and Roehl followed his lead, looking to succeed against the scourge of Africa where Ehrlich had gotten sidetracked onto syphilis. Soon hundreds of Bayer mice were infected with trypanosomes, the parasites that caused sleeping sickness, and scores of dyes were tested against the infection. Again the animal tests were critical. Simply mixing a potential medicine with a microorganism in a test tube was not good enough—it might kill the parasites or bacteria, but that was how you found antiseptics for washing your hands, not medicines that worked inside the body. Roehl ordered his test chemicals from the dyemakers at Leverkusen, who grumbled about the extra work. They had been hired to color fabrics, not to cure ailing mice. But Roehl was soon seeing interesting results where even Ehrlich had failed. There was nothing like a complete cure, not yet, and even when a dye seemed to slow the disease, there was another problem: The dyes sometimes colored the skin of the mice. A drug that tinted the patient orange or blue was not the best idea for the marketplace. So Roehl came up with what turned out to be an important idea. He thought that the curative ability of these compounds might lie not in their coloring ability but elsewhere in the chemical structure. In other words, it might be possible to separate the medicine from the color. He asked the Leverkusen chemists—to their initial confusion—to try to make him a sort of colorless dye.

It is a tribute to their skill that not only did they eventually understand what he was seeking but they found him one. While Roehl was away serving in the war in 1916, the chemists delivered to his laboratory for testing the 205th variation in the series they had been pursuing, an odorless, slightly bitter, baroquely complex, and entirely colorless derivative of urea. Roehl's well-trained assistants found that 205 could cure sleeping sickness in mice with few side effects. When he returned after the war, Roehl confirmed the results, but it was not until 1921 that Bayer was able to mount an expedition to South Africa, led by a distinguished member of the Prussian Institute for Infectious Diseases, to test its new substance. Sleeping sickness was the bane of colonial powers, afflicting tribespeople and colonists alike wherever they found the carrier of the parasites, the tsetse fly. The Nyanja, for instance, a tribe in the then-German colony of Cameroon, had been decimated by the disease, shrinking from twelve thousand tribespeople in 1914 to fewer than a thousand by 1922. It was impossible to make African colonies successful if there was sleeping sickness. So the German medical team trekked from Capetown to Rhodesia, the bearers hauling, with the expedition's tents, guns, and food, thirty kilos of Bayer 205. They mixed the medicine with rainwater, lined up hundreds of native patients, and gave them shots. Soon they were seeing what they reported as "biblical healings." Only three injections were needed to cure even the most advanced cases of sleeping sickness. In 1923 the new drug was packaged and released to the world as Germanin—a brand-name announcement that the world's best medical research was still, even after the war, found in Germany. Germanin opened up huge tracts of sub-Saharan Africa to European colonial ambitions. It is still one of the world's most effective anti–sleeping sickness medicines.

Roehl by then was hard at work finding the next big cure, this time for an even more widespread parasitic disease: malaria. This common fever afflicted more than half a billion people worldwide and killed somewhere between 2 and 4 million sufferers every year. There was already a treatment for malaria, of course, and had been for centuries: quinine, an extract from the bark of the cinchona tree in Peru. The problem was that access to natural quinine was jealously guarded. A myth among South American Indians where the tree grew naturally said that if it was stolen and grown elsewhere, they would die, so they sabotaged all attempts to take cuttings or seeds. The Dutch finally succeeded in buying some seeds from a British adventurer and started growing them on plantations in Java. The two sources—South American and Dutch—constituted almost the entire world's supply of quinine. Malaria was a problem not only in tropical countries but in warm, swampy land anywhere. The disease had been endemic in Rome and New Orleans until nearby swamps were drained. In the early 1900s, it continued to hamper trade and development in Asia, Africa, and South America. Everyone knew that a chemical alternative, a synthetic quinine, would be worth a fortune. That was why young Henry Perkin had searched for it back in the 1850s when he made mauve. Ehrlich, too, had tried and failed to find an antimalarial. During World War I, Germany's supply of quinine had dried up. The nation needed a synthetic alternative to break the Dutch and South American monopoly.

That was Roehl's next target. Before he could zero in on it, however, he had to find a suitable test animal. The malaria parasite, so lethal in man, could not be grown reliably in standard lab animals like mice or rabbits, but Roehl found that he could grow a closely related parasite in birds. He began to use canaries as his test animal for drug screening. Soon the Bayer animal facilities were full of canary cages. With a test animal in place, Roehl began following a trail that had started with Ehrlich's observation that his favorite blue dye, methylene blue, had shown a mild effect against the malaria parasite. Impressed by the Germanin success, the Bayer management team rewarded Roehl with expanded facilities for the antimalarial hunt. He was still part of a team, however, a central cog in a well-organized medical-research machine that now included another physician and two newly hired chemists, all dedicated to finding synthetic medicines. He no longer had to bother the dye chemists at Leverkusen to get his goods. Roehl's job was to suggest fruitful starting points (like methylene blue), then allow the chemists to tinker with the starter molecules, adding an atom or a side group here, subtracting one there, creating variations. The new chemicals were delivered to Roehl, who was in charge of testing and retesting them in test tubes and animals, throwing out those that did nothing and requesting new variations on those that had any sort of effect. If one variation worked better than another, the team asked why: What did this or that variation do to improve the effect? Then they would try more variations around the same theme, looking for something with a stronger effect. A team effort at this scale was something new in drug research—particularly in German research, where the tradition was to center a laboratory around a single brilliant, autocratic scientist (an Ehrlich, say). Hörlein's Bayer model was both larger and more decentralized. Roehl made suggestions, but he did not tell the chemists how to do their work. The chemists did not tell Roehl how to test. All results were sent up the line to administration for review and analysis. The Germans sometimes called it an American-style laboratory.

The size of the team, including all the assistants, animal keepers, and administrative support, translated into the ability to quickly screen hundreds of chemicals for medicinal abilities. Promising compounds now could be tested on several diseases at the same time: In the search that yielded Germanin, hundreds of chemicals were also tested on syphilis, in hopes of finding a replacement for the Salvarsan medicines (none was found). For every promising chemical, scores of others showed no effect whatsoever.

When the methylene blue trail played out, all those variations leading to another dead end, the Bayer researchers changed their focus to quinine. Chemical after chemical, all variations on the quinine pattern, were tested on cage after cage of canaries. It was years before Roehl's research group finally came up with something that worked: a synthetic chemical thirty times more effective against malaria than any other they had tested—and much more effective, gram for gram, than natural quinine itself. The Bayer team was not set up to test new drugs on humans, so they sent their find to the Institute for Tropical Diseases in Hamburg, where physicians determined that it worked well in malaria patients, especially when boosted with a dose of natural quinine. Field tests in malaria-infested areas followed; these were successful as well. In 1927, Bayer started marketing the drug under the trade name Plasmochin in Germany (Plasmoquine in the English-speaking world).

It was not perfect. There were side effects—side effects of some sort or another were now expected in all new drugs—and the chemical worked at only one stage during the complicated reproduction cycle of the malaria parasite, which meant that timing became critical for effective treatment. But, used correctly, it could cure. The world's first successful synthetic antimalarial, Plasmoquine broke the monopoly of the Dutch and South American growers. Bayer began marketing it aggressively, and money started to roll in. The company planned more expansion, new laboratories, a bigger drug-research program. Roehl was put in charge of parasitic (or, as they were also known, tropical) diseases, the area in which he had so much success. He never found a successful antibacterial chemical. Parasites and bacteria were far different organisms, so a completely new effort was planned to complement Roehl's tropical-disease unit, a group devoted to bacterial diseases only. In 1927, Domagk was hired to head it.

Domagk owed a great deal to Roehl. His first laboratory was a few rooms carved from Roehl's bustling operation, above a glassware storeroom. He saw how Roehl's system worked, how he ran his tests, how he correlated his results. Domagk inherited both Roehl's successful methods and a piece of the company support that started flowing to new-drug development thanks to the sales of Roehl's Germanin and Plasmoquine.

Still, it was not easy. There was Bayer Elberfeld itself, an aging factory along the Wupper River, in which Domagk now worked. Elberfeld had the brick-and-smokestacks look of the nineteenth century. It was vaporous, dark, and the river, in 1927 when the Domagks arrived, was foul-smelling from decades of chemical pollution. The factory huddled along the banks in a particularly steep part of the valley, a space so narrow that a suspension railway was built to freight in workers; it screeched and clanked all day long. Gerhard and Gertrude had grown up in the open fields and fresh winds of the lake country of the Brandenburg March. They hated the darkness, the smell, and the constriction. The people were different, too: The Domagks found that here in the Rhineland people did not seem as friendly as they had at home; they sometimes got the impression that the locals looked down on them.

It was not a good start, especially coupled with his small rooms above the storage area and the small amount of help he was given to start with, one technical assistant and two boys to wash glassware. Hörlein told him that the new laboratory was on the drawing board and would be completed soon. There would be more room and more help. He just needed to be patient.

Then tragedy struck. In 1929, Roehl, Bayer's drug-development star, a new Ehrlich well on his way to his own Nobel Prize, was hard at work finding an improved new version of his antimalarial. He was traveling in Egypt when he noticed a boil on his neck while he was shaving. As it turned out, the boil was infected with *Streptococcus,* the same germ that Sir Almroth Wright had identified as the most important cause of wound infections. The strep broke out of the boil and moved into his bloodstream. Roehl knew the physician's nomenclature for his condition: bacterial septicemia, an infection of the blood. This was a bacterial infection. There was no cure. Domagk had been looking for an antibacterial medicine for two years but had not yet found one. A few days later, Roehl was dead. He was forty-eight years old.

DOMAGK'S RESPONSE was to throw himself into his work. He was in a hurry to find cures, to ease suffering, and to make his name, but Hörlein took a longer view. When it came to industrial research, Hörlein was both a visionary and an optimist. By the time he hired Domagk, he knew that most researchers had given up on Ehrlich's magic bullets, convinced after years of fruitless searching that synthetic medicines were little more than a dream, one of Dr. Phantasmus's fantasies. In the two decades since Salvarsan, all the excitement had yielded only Germanin and Plasmoquine, which were effective (although, like Salvarsan, with occasionally severe side effects), but only against tropical diseases, parasitic illnesses that mattered most to relatively poor people halfway around the world. In Europe and North America, Bayer's major markets, most of the important infectious diseases were bacterial: pneumonia and tuberculosis, meningitis and blood infections, diphtheria and cholera. Chemical firms kept coming up with medicines they claimed would work, but none did. Ehrlich, it appeared, had been very, very lucky.

Hörlein was more patient than that. He wanted solid breakthroughs. He figured the search would take years. And he expected hundreds of failures along the way. But he had faith. One success against bacteria, just one, could open an entire field, lead them to a host of patentable drugs. IG Farben would reap enormous profits. Hörlein, with one foot in the research labs, the other in corporate boardrooms, approached the idea of synthetic medicines not from the standpoint of a single discipline—bacteriology, say, or chemistry, or pathology—but as the manager of an integrated business, able to devote enormous resources and the talents of many researchers to the quest. He kept his workers enthused and his bosses happy, led cheers when things looked hopeless and ensured a steady flow of money. As a former dye chemist, he also knew that the ways they had used to find new dyes—by taking a core compound and altering the bits around it—was the same as the process for finding new medicines. His team had proved the point with Germanin and Plasmoquine. This was Hörlein's greatest strength: the ability to build teams and systems, to ramp up the search for medicines, to expand drug research from the lab of a single scientist to an efficiently organized industrial process with carefully chosen specialists guided by a coordinated strategy.

Hörlein was also proudly German, and enthusiastic about the ways his company's success added to his nation's reputation. "Every Aspirin tablet or Salvarsan ampoule used in the most distant countries bears witness to the high position of German science and technology," he said in 1927. But his rhetoric belied the postwar realities: The demand for dyes had declined; international conflict had invigorated competing chemical industries in other nations; German pride had been wounded; the German economy was still in disarray. Hörlein believed, as did Duisberg, that one way to return Germany to the forefront was to find wonder medicines. IG Farben was ready to gamble. The giant firm had fat profits and massive reserves; money was available for big new investments. Synthetic gasoline was one of them. Synthetic rubber was another. And now, under Hörlein's guidance, synthetic medicine was given its chance.

In 1927, the funds Hörlein needed to expand the program began flowing. He designed his organization, hired Domagk, brought in new chemists, expanded the test-animal operation, broke apart old research teams, and built new ones. The former bacteriological laboratory was dissolved. A new pharmacologist was hired. Domagk's antibacterial-medicine program was launched. He started building new facilities, the most modern pharmaceutical laboratories in the world, the best equipped, the largest. The old Elberfeld chemical-research laboratories designed by Duisberg himself in the previous century—a warren of tiny lab spaces off a central corridor that the workers called "the Stables"—were kept as well, and there, up in the attic, was Workroom 4, a legendary lab space for chemists lit by skylights set in a low roof. Workroom 4 was old and somewhat cramped. It was unbearable in the summer, when the heat got so bad that the diethyl ether sometimes boiled in its bottles. But it was a badge of honor to work there. It was in Workroom 4 that chemists had created Germanin and Plasmoquine. It was a lucky place to work. And it was here that chemists started the flow of chemicals to be sent to Domagk for testing, in hopes of finding the next miracle: a drug that would stop bacteria.

HÖRLEIN'S PLAN for Domagk's new laboratory was a single three-story building divided into thirds, one end for biochemical research, the middle for tropical diseases, and the last third devoted to Domagk's research. Roehl's death had been a setback, but Bayer was bigger than any individual scientist. When Roehl died, the basic plan was retained, with a successor hired to take on the tropical-disease effort.

Domagk was happy. He adapted Roehl's (and Ehrlich's) system, overseeing the testing of compounds provided to him by chemists. Bayer had a crew of them in the Stables making chemicals for both Domagk's antibacterial effort and for the tropical-disease program. For maximum efficiency many of the same chemicals would be tested at the same time in both places. Soon Domagk was receiving new compounds from a half dozen chemists, the chemists pumping them out, Domagk testing them. They were relatively independent operations at Bayer. Only one of the chemists up in Workroom 4 was assigned specifically to Domagk. But one, as it turned out, was all he needed.

Josef Klarer was reputed to be something of a genius. Tall and handsome, a few years younger than Domagk, hired as part of the same 1927 expansion, Klarer had received his doctorate summa cum laude in Munich under the tutelage of Hans Fischer, himself a Nobel laureate. Observers had called Klarer's thesis on dye structures "sensational." He seemed destined for a stellar career in academia but had, it was said, turned down a professorship to come to Bayer. He was brilliant but also, as brilliant people often are, slightly unstable. Some chemists were theoretical, carefully thinking through structures, but clumsy in the lab; Klarer by contrast was a natural hands-on scientist with an inborn genius for lab work, a Mozart at the bench. Many chemists worked slowly and deliberately; Klarer was spontaneous and fast. He worked without any apparent plan, and he made it look easy. Yet there was something manic about him. At Bayer he could be found toiling fiercely at all hours. He ate irregularly, then disappeared for days at a time. He avoided talking with people and seemed gruff and touchy when forced into conversation. In the absence of communication, his coworkers gossiped: Klarer never slept; he had been severely wounded in the war; he had undergone a long convalescence (these last two were true). Most of his colleagues left him alone. The tradition at Bayer in any case was for chemists to work on their own, reporting up the ladder, not across the aisle to other chemists. It was a style suited to industrial secrecy. And it suited Klarer. The company appreciated his talents, so they allowed him to set his own work schedule, looked the other way when he took off, let him work all night when he needed to. Klarer made new molecules at a fantastic rate and sent them to Domagk for testing. He was the most productive chemist the company had. No one else could come close to his output.

The only person who could be even loosely be called Klarer's "friend" was Fritz Mietzsch, another Bayer chemist. They made an unlikely pair. Mietzsch, who had come to the company a few years earlier to work with Roehl, was a model of organization, a reserved man, a neat dresser who kept regular hours, a meticulous planner who outlined each coming week's work schedule in advance. Mietzsch was polite, quiet, scholarly, and respectful of authority. He never raised his voice. Mietzsch "kept his distance," as one writer put it, "which was also a characteristic trait of Domagk's." Klarer worked primarily for Domagk, Mietzsch for the tropical-disease unit, but their lab benches were close to each other, and he and Klarer bonded somehow, the older man appreciating Klarer's peculiar genius, the younger man turning to Mietzsch—himself a talented chemist, better grounded than Klarer in traditional techniques—when he ran into problems. There was a sense of the older chemist's keeping an eye on the mercurial younger man. That would become important later.

For now they worked quietly in Workroom 4, tracing the patterns of atoms, unlocking the structures of chemicals, finding ways to take them apart and re-form them, altering them slightly, creating new compounds, hoping that one would become Bayer's next miracle medicine.

CHAPTER EIGHT

DOMAGK COULD NOT TRY Klarer's chemicals against every kind of bacteria. There were too many that could cause disease—scores of major infections, hundreds of minor conditions. So for testing purposes he used a panel that included some of the most common and deadly: tuberculosis, pneumonia, *Staphylococcus, E. coli,* and the germ that killed Roehl: *Streptococcus pyogenes.*

Strep might seem an odd choice today when the only strep disease most people ever experience is a bad sore throat. In the 1920s, however, it was one of the most feared killers on earth. No one was safe from strep.

In the summer of 1924, a skinny, gawky teenage boy loped across the South Lawn of the White House with a tennis racket in his hand. He looked frail, but when he played tennis, he gave it everything he had. On this particular afternoon, he wished he'd worn socks under his sneakers. When he got back to his room after playing, he found a blister on his big toe, which he treated with iodine and forgot.

Two days later Calvin Coolidge Jr., son of the president of the United States, started to feel weak and feverish. He felt pain in his legs. He went to bed. When the White House physician, Admiral Joel Boone, examined the boy, he found the foot with the blister inflamed and tender. Boone cleaned the foot, bandaged it, and ordered a regimen of bed rest and regular changes of antiseptic dressings. When President Coolidge heard that his son was ill, he became unusually concerned. Calvin Jr. was his favorite, his youngest. Some historians later pointed out that the boy's face resembled that of the president's beloved mother, who had died young. Boone told the president and the First Lady not to be too concerned: The foot was infected, but with proper care healthy young men could shake off an infection like this within a week or two. He would keep a close eye on Calvin Jr.

The next day the boy's stiffness and pain were worse. Boone was a decisive physician—he had won a Medal of Honor in France during World War I—and he was not one to spend much time playing wait-and-see. He drew blood samples and drove them personally to Walter Reed Hospital. When he saw the results, his concern increased. It looked as if the boy's blood was infected with a particularly bad germ, a fast grower that was famous for spreading quickly through the body, shedding poison as it went. It looked like *Streptococcus pyogenes,* the main culprit in wound infections during World War I. Boone started moving quickly. When he got back to the White House, he found Calvin Jr.'s temperature soaring. He did what he could, eased the pain, cleaned the area, changed the dressings, and hoped the boy's immune response would kick in soon. That night he again consulted with the president and the First Lady. The next day, July 4, the president's birthday, the boy was worse. The planned festivities were toned down. Guests left the White House. The boy's room was turned into a hospital ward. Every medical expert on blood infections in the area was called in. The physicians tried everything they could to stop the infection, which was now moving up the boy's leg.

On July 5, news of Calvin Jr.'s condition hit the newspapers. Even the Republican president's political opponents rallied behind him: Attendees at the Democratic National Convention then under way in Madison Square Garden received regular bulletins on Calvin Jr.'s health and sent their sympathies to the White House. Phone calls and telegrams of support—more than ten thousand of them—began flooding in. The president seemed stunned, unable to concentrate on his work. He stopped what he was doing a dozen times that day and made his way to the boy's room. Journalist William Allen White told the story of the president's finding a rabbit in the White House garden, coaxing it close enough to catch, and trotting it up to his son's room. He got a smile in return. A Coolidge family friend told White that the president "would have carried him the whole of the White House grounds, a handful at a time, if it would have done any good."

Blood transfusions were sometimes tried in cases like this, although they were risky—blood-typing technology was crude, and the patient's own immune system could react against the transfusion. The next day, July 6, White House staffers began offering their blood. But by then Boone had decided to get the boy out of the White House. Calvin Jr. was taken in an ambulance to Walter Reed Hospital, with the pale First Lady following in a car behind. There, seven of the world's best physicians went to work, giving the boy injections of saline, blood transfusions, artificial respiration, a last-ditch operation. The chief justice of the Supreme Court wrote his wife, "The whole county is at the deathbed of young Calvin Coolidge." He then told her that he had talked with a physician who told him it had been hopeless from the start: Once a S. *pyogenes* infection hit the bloodstream, it was as bad as being bitten by a poisonous snake.

The president and the First Lady stayed at the hospital through the night while their son passed in and out of delirium. Toward the end he thought he was leading troops into battle and they were winning, a positive sign, his father thought. But then Calvin Jr.'s body went limp, and he murmured, "We surrender." Dr. Boone said forcefully, "No, Calvin, never surrender." The boy slipped into a coma. On July 7, five days after Boone first looked at a blister on the boy's toe, the president's son died. Calvin Coolidge fell into a deep depression that some historians believe marked his entire second term, turning the once-energetic politician into a man known to everyone afterward as "Silent Cal."

STREP WAS EVERY doctor's nightmare. The organisms could be found everywhere, in dirt and dust, in the human nose, on the skin, and in the throat. Most strains of strep were harmless. But a few were deadly, and when they got into the wrong place—beneath the skin, through a wound, into the blood—they could cause at least fifteen different human diseases, each so different from each other that in the 1920s researchers had still not untangled them. The worst strains of strep could secrete three poisons, wipe out red blood cells, raise fevers, eat through tissue, fight their way through the body's natural defenses, and create a bewildering variety of different diseases as they went. A strep-infected scratch could lead to the burning rash of erysipelas, the old St. Anthony's Fire; a bit deeper it became cellulitis, a potentially fatal infection of the subcutaneous tissue; if it got into the bloodstream, it caused septicemia, a blood infection; in the spinal fluid, meningitis. Some strep diseases were relatively mild, others vicious. There was no containing it. A sore throat, infected with strep, could progress to quinsy, a throat abscess that had to be opened and drained before it suffocated the patient, which could lead to "bull neck," with enormous gland swellings, often linked to a severe strep ear infection, sometimes a precursor to strep's breaking into the bloodstream or, worst of all, into the spinal cord, causing strep meningitis. Blood infections with this germ usually killed the patient; spinal fluid infections—strep meningitis—*always* did. Strep was responsible, they said, for half the white hairs on every physician's head.

By the time Domagk started working at Bayer, scientists had firmly identified strep as the cause of boils and abscesses, fevers and rashes, infections of wounds and of the heart, lungs, throat, blood, spinal cord, and middle ear. It had been identified as the cause of childbed fever. It was suspected—and later proved—to cause scarlet fever and rheumatic fever. Infections could start from something as minor as a pinprick, a sliver, or a burn; surgeons could die from a strep infection after nicking themselves during an operation. An artist had died of a strep infection after a model scratched his face. Worse, many of the diseases strep caused were capable of flashing into epidemics, passed on the hands, in nasal secretions, or through the saliva of carriers. Some strains could live for weeks in dust. If a deadly strain of strep took up residence in a hospital—and this happened all too often—it could be almost impossible to eradicate. Strep was a major reason doctors and nurses started wearing surgical masks. But even the most scrupulous precautions were often not enough. In 1930, the top four most serious hospital infections (those that threatened patients during hospital stays) were cellulitis, erysipelas, wound infections, and childbed fever. Every one was caused by strep. Taken together, strep diseases in Europe and North America alone during the 1920s were estimated to kill about 1.5 million victims a year. That same number adjusted for today's population would total more than current worldwide annual deaths from cholera, dysentery, typhoid, and AIDS combined.

Under the microscope all strains of strep look like twisted strings of beads (*streptós* is Greek for "twisted"; *coccus* is a round bacterium). Pasteur had tied them to human disease as far back as the 1870s. The earliest names given to the organisms were based on what they did: *S. pyogenes* meant pus-forming strep, *S. pneumoniae* was pneumonia-causing strep, *S. erysipelatus* caused erysipelas. As more diseases were tied to strep, it became clear that this was not a single killer but a clan of cousins, a rogue family within a much larger universe of mostly harmless strep. Each strain had a slightly different growth pattern, a slightly different set of identifying marks, a different way of causing disease. Sometimes a single disease could be caused by more than one member of the clan. Sometimes a single member of the clan could cause more than one disease. It was, to say the least, confusing.

And in the 1920s it made the only effective weapon they had for treating bacterial disease once it started in the human body—serum therapy—impossible. That was proved during an epidemic of pneumonia in army training camps in Texas. Nothing was helping the young soldiers, who were coming down with the disease by the hundreds. Many were dying. It got so bad that officials called in outside medical experts to figure out how to stop it. The scientists tried serum therapy—gathering samples of bacteria from the lungs of infected soldiers, shooting the strep into test animals, harvesting the anti-strep serum, shooting it into the soldiers—and failed. Any single serum was active against only a single strain of strep, and there appeared to be too many different strains at work in Texas. Like every bacterial epidemic, this one eventually flared out on its own. The pressure was off, but the researchers kept trying to figure out why their efforts had failed. To untangle the strep mess, they hired a bright young Wellesley woman, a former French major who switched to microbiology when her roommate told her how interesting the classes were. Her name was Rebecca Craighill, but she became famous in the annals of science under her married name, Rebecca Lancefield. The switch from romance language to research lab was a lucky one for science. Lancefield was happy to get a job as a lab assistant and happy to take on what a more experienced researcher might have seen as the impossible task of unknotting the relationships within the strep family. She turned out to be a phenomenon—tireless, skillful with her hands, patient, meticulous, and insightful. She figured out how to separate different strains of strep, how to grow them in the lab, and how to use the antibodies of animals as exquisite sensors able to differentiate minute variances between one kind of strep and another. She would isolate a strain of strep, grow it in quantity, then puree the bacteria and inject the resulting soup into animals. The injection was not an infection—the puree contained no live bacteria, only bits and pieces—but the animal would still sense the injected material as foreign and mobilize its immune system to fight it, producing antibodies highly specific to the particulars of the invading substance. If one strain of strep differed even slightly from the last, the antibodies would sense it. Lancefield harvested the antibody-containing blood, spun it down in a centrifuge, and collected the yellow serum. Each serum was matched precisely to a single strain of strep. She slowly built a library of serums that she could use as highly specific probes. If the serum responded strongly to an unknown strep, then the unknown strain was closely related to the one used to make the serum. If the reaction was weaker, they were not closely related. She used the serum probes to explore the world of strep and identify its members with pinpoint accuracy.

Through the 1920s she found that the strep causing erysipelas was slightly different from the one causing childbed fever, which was slightly different from the one causing meningitis, and so on. For a time she studied "green" strep, which some people thought caused rheumatic fever. Her work helped to show that green strep was a large and varied family of related strains, mostly harmless (although some green strep could cause a form of heart infection).

Her real passion was the study of another large group of strep, the most dangerous, the strains that caused most strep diseases—the ones that killed Calvin Coolidge Jr. Once again the bacterial world proved far more complex than anyone had thought. These disease-causing strep were part of a larger group called hemolytic strep, named for their ability to destroy red blood cells. Lancefield discovered that hemolytic strep could be divided into three large subgroups, which she named simply alpha, beta, and gamma. Not every hemolytic strep was a danger to human health, however. The alphas and gammas were relatively harmless. It was the beta hemolytic strep that included most of the killers. She studied the betas intensively, breaking them into subgroups—A, B, C, all the way through O—based on differences detected by antibodies. Then she focused on Group A, where again the worst killers seemed to be concentrated. She identified many of them, but she never did find them all. Researchers eventually discovered more than forty separate strains of Group A beta hemolytic streps, each different enough from the others to cause unique diseases and reactions. A serum raised against one did not work very effectively against another. No wonder serum therapy had not worked in Texas. No wonder Sir Almroth Wright's serum had failed to stop strep-caused wound infections in Boulogne.

Lancefield's results, her list of strep types and subtypes lengthening through the 1920s, offered little hope that any medicine would ever be found to stop them. There were just too many types.

Lancefield's findings also made life more difficult for Domagk. Strep, he knew from the work of Sir Almroth Wright and others, were the single most important cause of the wound infections that had given him his mission in life. If strep existed in so many forms, each a bit different from the others, what were the chances of finding a single chemical that would work against all of them? He faced a more immediate practical problem as well. Domagk wanted to include strep in his test panel of germs, the group of microbes he used to screen chemicals at Bayer, but he could not economically test every chemical against forty strains of strep, or twenty, or even five. He needed a single strain, one that could kill both mice and men, a superstrep with which he could reliably infect test animals.

"Reliably," for his purposes, meant a strep that, when injected into a mouse, always grew, always spread quickly, and always killed. Mice were his test animals of choice, because they reproduced rapidly and could be bred into uniform strains, lessening mouse-to-mouse variations that could throw off his results. He did not want simply to make his mice sick. He wanted them dead. Within a few days was best. Death was definite and unarguable, a certain end point for a test, a plus or minus in his lab book, a precise point in time; any lab assistant could assess it without any doubt or personal opinion. That made for reliable science.

He began sampling and isolating strep from human patients and found that most strains did not kill mice the way he wanted. Strep, in addition to their other qualities, were fastidious germs, able to survive in a number of environments but flourish in few. Most of the strains of interest to health workers were so well adapted to the human body that the temperature and composition of growth broth had to closely mimic that of human blood. Domagk's lab used an egg- or meat-infusion bouillion containing serum or blood and 0.1 percent glucose, in an atmosphere ranging from 5 to 10 percent carbon dioxide. Any deviation and the microbes refused to grow. Domagk learned the tricks.

Then he went looking for his superstrep. He made contacts in local hospitals, asked physicians to keep an eye open for particularly vicious strep infections, collected samples of bacteria from the bodies of the dead. The strep he gathered varied terrifically in their ability to kill mice. None seemed to provide consistent results over time. It was months before he isolated a strain of strep from a patient who had died from a particularly aggressive case of blood poisoning. This particular strep was so powerful that a fraction of a drop from a bacterial culture diluted to one part in one hundred thousand killed every mouse he gave it to within two or three days. Most died within twenty-four hours. This superstrep might not be representative of all strep—what strain was?—but at least it provided Domagk with a reliable germ to use in animal tests, a peerless killer. It was the best he could do.

Test system in place, he made anti-strep experiments part of his standard disease panel. By 1929, when he moved into the newly completed laboratory Hörlein had promised, Domagk had made a fine art of animal testing. It was now possible for him, his growing group of six assistants (all of them women), and his animal-care support staff to thoroughly test more than thirty new chemicals per week, both in vitro (in test tubes) and in vivo (in animals), delivered three different ways (intravenously, subcutaneously, and by mouth), plus the occasional addition of a nonfatal disease like gonorrhea or other germ of interest, plus tests against cancer cells. Not every new chemical sent to Domagk's lab got the full battery—some were given fewer tests for initial screening, and on some days a particular culture of germs might not be ready for injection—but in every case every chemical was tested against a variety of diseases in living animals. It was just what Hörlein wanted: a smooth-functioning, reliable machine for discovery. He signed Domagk to a permanent contract. The months went by, the chemicals came in for testing, the mice died in batches of six or ten at a time, thousands upon thousands as the years passed. There were no quick discoveries, but they did not expect that. Domagk perfected and refined his system. Even the dead animals had their function. Domagk did postmortem exams once a week, running his scalpel down each animal's abdomen, peeling back the skin and muscle, examining the organs, looking for swelling, discoloration, abscesses. Then he took fluid samples and removed organs, carefully sliced paper-thin sections, stained and mounted his samples on slides, examined them under the microscope, gauged where and how the bacteria had caused death. He trusted this work to no one else, making every exam of every animal himself. During these times he shut himself off and refused visitors and phone calls. Domagk's system raised the testing of prospective drugs to a new level of precision and a vastly expanded scale. By 1929, that system was working at peak capacity.

The only thing missing was a positive result.

CHAPTER NINE

Sir Almroth Wright's research group had found strep involved in roughly three-quarters of all wound infections in the hospital at Boulogne. It was the primary cause of the most grievous cases, including gas gangrene, where the strep acted as a first wave of infection. There was strep everywhere at Boulogne; the germs were blown into wounds along with the dirt and torn bits of uniforms, grew in tissue easily, invaded the body readily, weakened the patients with poisons, and soaked up oxygen in the wound. Sir Almroth had learned everything there was to know about strep wound infections—except how to stop them.

Now, in the late 1920s, he had grown dispirited and increasingly cranky. He had thought that serum therapies or vaccines would stop wound infections. He had been proved wrong. But he was also certain that drugs would never solve the problem. That had been proved to his satisfaction back in 1911 during his ill-fated trip to South Africa. The Optochin disaster had turned him forever against the Germans and their fascination with using laboratory chemicals to cure disease, a quixotic endeavor, Sir Almroth thought, a field of failure that was now being called chemotherapy (used today only in relation to cancer treatment, the word "chemotherapy" in the 1920s referred to treating any infectious disease with chemicals).

Leonard Colebrook was not so sure. Sir Almroth's second-in-command at the St. Mary's Inoculation Department in London—the scientists that called themselves the House of Lords—Colebrook had been with Sir Almroth in both the diamond mines and the casino at Boulogne. He had seen the same horrors, the blinded miners, the soldiers with yards of rubber tubing stuck into their abdomens, the wounds bathed in bleach. He had heard the men crying out as their injuries were treated. Colebrook had always done whatever Sir Almroth asked, but now he was beginning to grow weary of the Old Man's constant railing against chemical cures. Colebrook quietly and respectfully disagreed. Optochin might have been a fiasco, but Colebrook placed more emphasis on the fact that Ehrlich's Salvarsan had been a success. Salvarsan worked against only one disease, it was true, syphilis, an unusual malady caused by an unusual germ, and even then only with excruciating side effects. But it worked. To Sir Almroth, Salvarsan was an anomaly. To Colebrook, Salvarsan was a signpost. It proved that chemicals could stop disease, and it pointed the way to more cures. All of the failures that followed—the chemicals that cured mice but not people, the chemicals that made patients sicker than their disease did—failed to dissuade him. Many researchers in the 1920s gave up looking for drugs to cure infectious disease. But Colebrook—like Hörlein and Domagk—continued to believe.

Belief was central to Colebrook's nature. He was profoundly Christian. He did not proselytize or criticize those who believed differently, but he did his best in daily life to demonstrate the power of Christ's teachings. He was hardworking yet softhearted, ambitious yet supportive of others, authoritative yet approachable. He sometimes even let slip a bone-dry sense of humor. His was the science of service. Medicine was a way to help the afflicted. As a student he had even planned to become a medical missionary, perhaps in Africa or Asia, bringing the light of modern science and the teachings of the Church of England to the darker parts of the world. Although he had ended up instead in St. Mary's working for Sir Almroth, Colebrook's idealism and missionary zeal still showed. He worked long hours, did not care much about salary, and did not treat his patients like experimental subjects. Many of those admitted to St. Mary's were women who had fallen ill in childbirth, all too common in hospitals at the time; often the new mothers' concerns about their families added to their physical ills. Colebrook sat up with them, held their hands, listened, counseled, and consoled. Sometimes he spent all night in the wards. He was never considered brilliant, but he was the sort of doctor everyone wanted.

And he was good in the laboratory. Sir Almroth, an unashamed atheist, put up with Colebrook's faith in part because Colebrook was a strong, reliable researcher and in part because Colebrook was as close as he would ever come to having a son. They were very different but complemented each other. Wright was combative and outspoken; Colebrook was quiet and restrained. Wright was certain his theories were correct; Colebrook doubted any theory until it was proved and then re-proved. Wright was perfectly capable of exaggerating to make a point; Colebrook was described by a coworker as "profoundly honest." They made a good team. Sir Almroth's energy helped Colebrook come alive; the older man grounded the younger in the meticulous techniques required by modern medical research and raised his ambition to take on the Big Questions, the ones that counted in medical history. Wright gave Colebrook some backbone and drive. Colebrook, in turn, was Sir Almroth's faithful second-in-command, loyal supporter, and reliable advocate.

Colebrook's honesty and diligence were much needed in the lab. The House of Lords was a tight, productive group, but they were mostly young men and they sometimes acted it. They all had nicknames (Colebrook's was "Coli"). They all lived close to St. Mary's so they could be available twenty-four hours a day to monitor experiments. Getting married was looked upon with suspicion, although Colebrook did marry at the beginning of the war. They worked hard, put in long hours, and occasionally played: Alexander Fleming, for instance, developed a talent for spreading patterns of different bacteria on agar plates, which after a few days would blossom into a colorful picture of, say, a ballerina. Most of them followed the Old Man to Boulogne and learned the same basic lesson: Once a wound infection started, little could be done to stop it. Certainly no drug could do anything.

After they returned to St. Mary's at the war's end, the reconstituted House of Lords went back to studying inoculations, vaccines, and immunity. Quietly through the 1920s, however, so as not to disturb the Old Man, Leonard Colebrook pursued what would have seemed in that environment an almost illicit interest in chemical therapies. He read widely, especially the German journals, and then began his own experiments, following up on techniques Sir Almroth's group had developed to fight wound infections in the casino. The first substances he tested were arsenicals, chemicals related to Salvarsan.

He immediately faced a problem. At Boulogne they had unlimited numbers of wounded on whom to test their ideas. With the war over, Colebrook faced a limiting factor at St. Mary's: There were not enough wound infections to support large-scale drug tests. So Colebrook turned to a different kind of infected wound, a wound found in peacetime, a wound that killed young women: the wound of childbirth.

In the 1920s the paradigm for obstetrics—a field that primarily male physicians had finally taken over, during the previous three centuries, from primarily female midwives—was that of illness. "Pregnancy is a disease of nine months' duration," one physician had quipped; another advised, "It is best to consider every labor case as a severe operation." Their remarks underscored the pessimism of caregivers who lost many new mothers after childbirth. The process of birth included a natural wound, deep in the mother's body, where the placenta detached from the uterus. Most of the time, it healed without a problem. Like every wound, however, it could offer a route for infection. Every new mother, a popular magazine of the day explained, was in an "exquisitely susceptible, wounded condition." That was a basic reason physicians of the day wanted deliveries done not at home but in hospitals, where new mothers could get adequate care and time for recovery.

Hospitals themselves, unfortunately, were often centers for infection. New mothers—especially those in maternity wards—risked a disease called childbed fever, endemic in many hospitals, that killed tens of thousands of women every year. Colebrook saw the results at St. Mary's, the dying mothers, the shattered families. He determined to do everything he could to stop it.

His first studies showed that childbed fever was caused by the same strains of *Streptococcus* that had been found in soldiers in Boulogne, the strep that Sir Almroth's research had pinpointed as the primary cause of wound infections. Following Sir Almroth's lead, Colebrook first tried anti-strep vaccines to prevent childbed fever (which failed, just as they had failed the soldiers in France), and then turned to chemicals. Using newly delivered mothers at St. Mary's and other maternity hospitals as his test group, he mounted an extensive series of experiments gauging the effects of arsenic-containing medicines.

Childbed fever was not a new disease (Hippocrates had described a case in ancient Greece), but it had been no more than an occasional problem back when most women were delivered at home. In the seventeenth century, however, what had been a scatter of isolated cases turned into a horrifying epidemic. It showed up first in Paris, in the Hôtel Dieu (God's Hostel), the city's largest and poorest hospital. Founded originally as a wing of Notre Dame, the hospital by the early 1600s was a demonstration of the dangers of mixing pure Christian charity with unbridled municipal growth. The problem was a combination of the hospital's underlying mission of offering succor to the poor—no one, no matter how poor, was turned away from the Hôtel Dieu—and the ballooning number of poor needing to be saved. By the seventeenth century, the Hôtel Dieu was drastically overcrowded. It sprawled over both sides of the Seine, the wards connected by a bridge devoted to the hospital's use. It was here, in a two-story building built on the bridge itself, that the world's first maternity ward was created in 1626. It was followed twenty years later by the world's first epidemic of childbed fever. It all started as a very good idea: Separating new mothers from the sick and wounded in the rest of the wards represented an advance in medical practice that became a model for other hospitals. "Lying-in hospitals," as they were often called, sprang up in leading cities of Europe and the Americas during the next two centuries. Unfortunately, given the poverty of the institution and the unfortunates it served, the lying-in ward at the Hôtel Dieu was primitive even by the medical standards of the seventeenth century. Patients were packed into oversize beds called *grands lits*—two, four, even six to a bed, head to foot. Pregnant women, many of them prostitutes, all of them penniless (those with money tended to avoid hospitals), would arrive at the door of the Hôtel Dieu late in their pregnancies, often at the point of giving birth. They would undergo a cursory examination before being sent to the new ward over the river to await delivery. Those awaiting birth were mixed with those who had already delivered. The women slept with their babies. It was not uncommon for infants to smother when women rolled over in their sleep. Every day the senior doctors would arrive on their rounds followed closely by a gaggle of students. They would pull the women's covers down, pass hands over their abdomens, point, prod, and discuss. Although the physicians' wigs were carefully powdered, their hands were generally unwashed. Christian care, which emphasized purity of the soul over that of the body, had replaced Roman hygiene with frequent prayers and infrequent baths. In Paris the privies and slaughterhouses (as well as the hospital wards of the Hôtel Dieu) dumped their waste into the Seine, then drew drinking and washing water from the same source. Bedding was washed infrequently. Lice and fleas abounded. There were no operating rooms, anesthesia did not exist, and physicians often performed surgery in the middle of the wards while the other patients in their beds looked on.

The new mothers who came down with childbed fever felt it first a day or two after delivery, with a looseness in the bowels and stomach pain. The disease progressed very quickly. Within hours the physicians might note a drying and hardening of the uterus, excruciating pain, headache, sometimes a cough, then a high fever. Occasionally the tongue turned black. In other cases the abdomen became distended, almost like a second pregnancy, the skin tight as a drum, so sensitive that women could not stand to have anything touch their bodies, not even blankets. At the Hôtel Dieu, little attempt was made to separate the sick women from those still healthy; the groups were simply placed on opposite sides of the same large room. The victims moaned and screamed. Healthy women awaiting delivery on the other side of the room sometimes became wildly alarmed (an understandable reaction that French physicians of the day classified as *hystérique,* an emotional outburst linked to a derangement of the uterus, a condition they believed might add to the risk of developing childbed fever). All the usual remedies were tried on the sick women: doses of strong laxatives to empty the stomach and bowels; copious bleeding, cupping, and leeching; even doses of the miraculous Countess's Powder (a pulverized tree bark from the New World that sometimes cured fevers, a medicine later called quinine). Nothing worked. The victims could do little but pray to St. Margaret, the patron saint of childbirth. The physicians eased their pain with opium preparations, provided what comfort they could, watched, and waited. Some of the women recovered. Most of them did not. The last hours were never pleasant. Death was seen as a mercy.

The first epidemic that raged through God's Hostel in 1646 killed scores of new mothers within a few weeks. It even afflicted some of those attending the women, nuns included, who came down with a fatal fever that seemed in some ways similar. There was great alarm, but the poorest pregnant women of Paris kept coming, deciding that the once-in-a-lifetime chance for weeks of bed rest and hot food outweighed the risk of disease. The hospital's premier president believed that the problem was related to location: Below the second-floor women's ward on the bridge was a room where open wounds were being treated, wounds that sometimes putrefied, releasing noxious vapors. Foul air rising from below, he believed, carried this disease to the women, miasma arising from rotting flesh. Confirming evidence of this theory was found when the dead mothers were autopsied. When physicians cut open the bodies of the newly dead women, there arose such a terrible smell of putrefaction that attending students sometimes fainted. Other physicians thought that rather than miasma the disease might be caused by rotting bits of leftover placenta or other tissues remaining in the uterus, the detritus of delivery. Others viewing the postmortems on the diseased women found on the surface of the intestines what appeared to be curdled milk, leading to a theory that the new disease somehow had been caused by "milk metastasis" related to lactation in the new mothers, a sort of cancer of the milk.

No theory explained it fully, and certainly no treatment worked against it. Then, as suddenly as it had arrived, the epidemic of childbed fever at the Hôtel Dieu disappeared. It was not gone for good, however. It returned a few years later, flared and disappeared, then flared again, more often as time went on. Soon it became an annual visitor to Paris, the number of cases rising in the winter months and subsiding in the summer.

And then the disease began to travel. An epidemic hit the lying-in hospital in Lyons in 1750, then London in 1760, and Dublin in 1763. Childbed fever quickly became a worldwide epidemic, spreading east to Vienna and west to the United States; at its peak in 1772, it was killing up to one in every five new mothers. During an outbreak in 1773 at the Royal Infirmary of Edinburgh in Scotland, almost every woman who delivered a baby was seized by childbed fever, and all of them who caught the disease died. But in Edinburgh the physician in charge of the maternity wards, a Dr. Young, fought back. He decided to treat childbed fever as if it were the plague. After seeing six mothers die in quick succession, he cleared and shut the maternity wards; ordered the mattresses, pillows, and blankets torn off the beds and burned; filled the wards with smoke to rid them of corrupted air; threw open the windows during the day to air the rooms; and shot off gunpowder wherever there had been disease. His actions were based on the old idea of miasma; he was dispelling the bad air. When he felt that the wards had been purified, he ordered every surface in the rooms washed and the walls in the rooms and halls repainted. New bedding was ordered. Then he brought back his patients. And the epidemic was gone. His work was ignored for a while, but by the mid-nineteenth century Young's approach was becoming widely employed. Infected wards would often be closed and cleansed, driving away the fever for a time. But it always returned. If it got too bad, hospital governing boards were known to recommend simply tearing down or burning the wards and rebuilding.

There were still no answers, only questions: Why newly delivered mothers? Why were some hospitals decimated while others in the same city, in the same winter, rarely saw a case? Within a single hospital, the disease might be rampant in one ward, a minor problem in another. Within a single ward, some women fell prey, and others never caught it. Of those who caught it, some recovered, while others died. It was especially galling that epidemics struck hospitals, the centers of the most advanced care, while the disease was rarely a problem for midwives delivering babies at even the filthiest private homes.

Two centuries after the first outbreak at the Hôtel Dieu, physicians seemed no closer to finding either the cause or cure of childbed fever, although theories continued to sprout: Some believed it was due to "autogenesis," a disease the women brought upon themselves through inadequate hygiene; others subscribed to the theory of "crasis," the spontaneous appearance of contagion in the blood; still others linked it to errors in diet, inadequate sewer systems, the shame of being an unmarried mother, or that convenient catchall for many female problems: hysteria. At least there was a solution for that: If a newly delivered woman seemed unusually nervous or irritable, one veteran doctor advised that she be dosed with laudanum to "restore rest to the body and tranquility to the mind."

Childbed fever caused the "utmost alarm in the physician," noted one nineteenth-century observer, sensitive to the psychological damage physicians themselves could undergo after repeatedly experiencing "the anxiety and anguish of those so lately rejoicing, the blighting of the sweetest hopes in life, and finally the rupture of its dearest ties, and the melancholy desolation of a home but lately the abode of happiness." Around 1840, after attending forty-five cases of childbed fever in a single year, Dr. Rutter of Philadelphia finally broke under the strain, leaving the city, burning his clothes, shaving his head, beard, and mustache, paring his nails to the quick, and scrubbing himself relentlessly in hopes of somehow protecting his patients. Unfortunately, when he returned to practice, the next new mother he attended died from the fever.

Cases like Dr. Rutter's led to new ideas about the source of the disease, put forward by a brilliant young American named Oliver Wendell Holmes. Known today primarily as the father of a future U.S. Supreme Court justice, Holmes in the 1840s was known as a wunderkind: a skilled physician who in his early thirties was already teaching anatomy and physiology at Harvard, simultaneously excelling not only in medicine but in literature, author of the popular poem "Old Ironsides" at age twenty-one, a regular contributor to literary magazines ever since. He was an essayist, a theorist, a practitioner, and a wide-ranging thinker. What Holmes had in abundance was outrage: outrage at the depredations of a disease that he believed could be stopped and outrage at the way in which his medical colleagues were ignoring the facts. In 1843, he published a blistering critique called the "Contagiousness of Puerperal Fever" (the medical name for childbed fever), presenting case after case in which it was apparent that the disease demonstrated its own sort of logic, occuring in clusters linked by geography and personal contact. His most noteworthy idea was that the disease could be carried from victim to victim by physicians and nurses. He wrote passionately and persuasively, making his case not only with numbers but with well-told stories: for instance, the tale of the doctor who, after assisting in the postmortem of a new mother who had died of childbed fever, carried her pelvic viscera in his pocket while attending another birth later that evening, followed by the death of the second woman a few days later, followed by the death of the woman he delivered the next morning, followed by the deaths of many of his other patients during the next few weeks. Holmes's essay consisted, in fact, more of anecdotes than statistically significant data, but its cumulative impact—case after case where groups of dying mothers were attended by the same physician, while another physician in the same town might deliver many babies without seeing the disease—was overwhelming. "In the view of these facts it does appear a singular coincidence that one man or woman should have ten, twenty, thirty, or seventy cases of this rare disease following his or her footsteps with the keenness of a beagle, through the streets and lanes of a crowded city, while the scores that cross the same paths on the same errands know it only by name," he wrote.

Holmes then suggested some commonsense precautions to lower the incidence of childbed fever: improved washing, especially after attending an autopsy; avoiding new patients for a few weeks after attending a single case of childbed fever; completely shutting down an obstetrical practice for a month if the physician was linked to two cases in short order. Then he added some bite to his conclusion: "The time has come when the existence of a private pestilence in the sphere of a single physician should be looked upon not as a misfortune, but a crime."

There was a natural reaction from physicians far more experienced in the field than the young Holmes, who was, after all, less an obstetrician than he was a poet. Charles Meigs, a leading obstetrician and teacher of the craft, a man who believed that childbed fever was spread according to the mysteries of God's Providence, was incensed at the idea that he or his colleagues might be somehow responsible. "Doctors are gentlemen," he maintained, "and gentlemen's hands are clean."

Holmes had made an impact, but more than stories would be needed to persuade the medical establishment to change its practices.

VIENNA'S LYING-IN HOSPITAL, its great Allgemeines Krankenhaus, was in the mid-1800s one of the world's largest maternity facilities, an enormous complex, a paradigm of cleanliness and quality. There, a minor member of the staff, a balding, excitable obstetrician of Hungarian extraction named Ignaz Semmelweis, also became interested in childbed fever. At the Allgemeines Krankenhaus, infants were delivered in two different areas of the hospital: Division 1, attended primarily by professors and medical students, and Division 2, staffed mainly by midwives. It was widely believed by the women of Vienna (and by some of the medical staff, too) that if you were to deliver at the Allgemeines Krankenhaus, you were best advised to avoid Division 1, the students' area, which was well known to be rife with childbed fever, and instead do everything you could to be cared for in the much safer midwives' area. Semmelweis devoted himself to finding out why. Meticulous records were kept at the Allgemeines Krankenhaus, and Semmelweis pored over them. He found that what was rumored to be true was in fact true: Between 1841 and 1856, new mothers had died in Division 1 at a rate three times higher than in Division 2. The overwhelming cause was childbed fever. The usual explanations for the spread of the disease—overcrowding, putridity of the atmosphere, the stresses of unwed motherhood—did not serve to explain the difference, Semmelweis found. Generally, all those conditions were worse in the midwives' area than in the students'. He began autopsying the Viennese victims of childbed fever, looking for the killers. The critical observation came when one of his colleagues, Dr. Jakob Kolletschka, cut his hand during a postmortem exam of a childbed fever victim. Before he died a few days later, Kolletschka fell ill with fever, headache, and abdominal pain, symptoms similar to those suffered by the childbed fever victims he had been autopsying. Semmelweis put two and two together. Students at the Vienna Lying-In Hospital routinely attended postmortems of women who had died from childbed fever, then walked into the labor wards and assisted with vaginal exams and deliveries. Childbed fever was rampant in the wards attended by students. Something appeared to be passing the infection from the dead women to the students and physicians, then from them to women in the delivery room. Semmelweis thought it likely, given the pattern, that the student physicians were carrying something, some bits of infectious material, perhaps scraps of tissue from the autopsies, from the autopsy area to the wards. These were the days before Lister and before Pasteur had shown that bacteria could cause disease. Hand washing was minimal, if it was done at all. The students and medical staff wore no gloves. All the doctors, young and old, generally wore the same clothes for days. Semmelweis came to believe that the infectious material was likely carried on the hands. To test the idea, he instructed all of his students to start washing their hands thoroughly in chloride of lime after any autopsy and before touching any patient. Then he tracked the results. As he had hoped, deaths among women treated by his students fell dramatically. Semmelweis excitedly told everyone about his findings, and soon his hand-washing practices were adopted throughout the Vienna Lying-In Hospital. Within a few years, the death rate in Division 1 fell to match that in Division 2. The pregnant women of Vienna stopped begging to be given a midwife. Semmelweis published his results in 1861. "Puerperal fever is caused by conveyance to the pregnant woman of putrid particles derived from living organisms, through the agency of the examining fingers," he wrote. "Consequently must I make my confession that God only knows the number of women whom I have consigned prematurely to the grave."

Feeling the burden of guilt, he felt compelled to spread that feeling to his colleagues. He began calling his fellow physicians "murderers" and "medical Neros" if they refused to take his advice, came under attack for his intemperance, and fell into a deep depression. In 1865, he was committed to a psychiatric institution in Vienna. Two weeks later he died—likely from blood poisoning following a cut to his finger.

Semmelweis's methods were soon adopted by lying-in hospitals around the world. He had helped to control the spread of childbed fever. But he had not found precisely what caused it—and without a known cause, there would be no cure. Fewer new mothers came down with childbed fever after Semmelweis, but once a woman did contract the disease, her chances of living through it were no better than they were before. In a good year, one new mother in twenty who was infected with the disease would die. In a bad year, in an epidemic year, childbed fever could still kill up to one in four infected women.

By the time Leonard Colebrook turned his attention to the disease in the 1920s, it was responsible for the deaths of tens of thousands of new mothers every year, and he was becoming increasingly convinced that the only way to stop it was through treatment with chemicals. In 1928, he spent the summer in a German laboratory, honing both his chemical skills and his knowledge of the language, a must for keeping up with the fundamental work coming out of the country. By 1929, Colebrook had earned a reputation as England's leading expert in the use of chemicals to fight bacteria. Through the late 1920s and into the next decade, Colebrook devoted himself to the fight against childbed fever. During those years Colebrook could often be seen in his open Morris Oxford car, macintosh cinched tight, traveling from hospital to hospital, Clapham to Hampstead, tending sick women, consoling, seeking a cure.

Arsenicals, it soon became clear, were not it. Colebrook's attempt to cure using the noxious chemicals was more than unpromising. It was a "total failure," one colleague noted; more than a hundred women were treated with the harsh drugs without a single cure. Arsenicals proved not only ineffective but often toxic in childbed fever patients, weakening the women, making them nauseous. Nothing seemed to work. Things were so bad by 1930 that one leading physician in the field told a colleague that he was sometimes treating his childbed fever patients with intravenous alcohol. "When I protested that all this would do would be to make them drunk," the second physician wrote, "he agreed with me, but smiled and added that it so improved their morale that they became better able to deal with the infections themselves."

Disappointed after his failure with arsenicals, Colebrook gave up not only his chemotherapeutic experiments for several years but every form of attempted cure, stopped giving transfusions and serum therapy, and decided that the best thing to do for new mothers was to provide them with the best nursing care possible and otherwise leave them alone. At least that way he would do no harm.

Then came a new opportunity. In 1931, he took a job on his own, overseeing a unit at Queen Charlotte's, a brand-new maternity hospital on Goldhawk Road. Queen Charlotte's included a hospital-within-a-hospital devoted to the isolation, care, and study of childbed fever victims, with an attached laboratory. Colebrook was asked to run it. It was a chance, finally, at age forty-eight, to leave Sir Almroth's laboratory and prove himself.

He made the most of it. Queen Charlotte's was modern, well equipped, almost luxurious compared to the dark, Victorian-era St. Mary's. His childbed fever unit, which they called the "Isolation Block," contained more than thirty beds; there was an abundance of natural light and balconies overlooking a garden. The research laboratory, equipped according to Colebrook's precise instructions, was up-to-date and efficient.

Streptococcus was now known to be the cause of childbed fever, and Colebrook did his best to keep his unit from spreading it, insisting from the beginning that the Isolation Block be a model of modern hygiene. Unhappy with soap and water, he studied all available antiseptics to find the best possible treatments for washing hands, trying them on himself by first smearing his hands with live strep (always checking first to make sure that he had no scratches or pinpricks in the skin), then scrubbing thoroughly with whatever the chemical wash of the day was, and running tests to see how many bacteria survived. His work showed that Dettol, a little-known antiseptic, was the best anti-strep wash, and his encouragement ensured its wide use not only throughout Queen Charlotte's but in other hospitals. Colebrook dictated that all Isolation Block nurses and doctors wear masks and gloves at all times when treating patients and made certain that all instruments were thoroughly sterilized. He even designed the door handles in his unit so they could be opened without the use of hands. No one was sure exactly where or how long strep could live outside the body. Until it was known, Colebrook behaved as if they could live everywhere, even on door handles. His rules were strict, but he made certain they were enforced, eventually convincing even the most hide-bound older staff members to meet his high standards.

He and his younger sister Dora, a bacteriologist, began a series of experiments to track the strep that caused childbed fever. How did the disease get started in a new facility? How did it spread? The general belief at the time was that the patients themselves carried it in, that it was the strep in their vaginas or borne in their own noses or throats that led to the disease. Using the strep-typing techniques that Lancefield pioneered, the Colebrooks followed the trail of cases, sampled the people involved, and demonstrated that two-thirds of the time the deadly infections started with strep from caregivers—nurses and doctors who carried it (usually in their noses) without symptoms—asymptomatic carriers, they were called. There were many asymptomatic carriers in a typical hospital. There were many strep carriers everywhere, researchers found—roughly one in ten people carry potentially deadly strep strains in their nose or throat without knowing it.

After that, everyone who wanted to be hired into the Isolation Block had first to undergo testing to see if he or she was a strep carrier. Colebrook recommended that everyone at Queen Charlotte's who came into contact with pregnant or newly delivered women be tested. He was a model for his team, working tirelessly treating patients during the day, often returning in the evenings to talk with them, easing their fears, boosting morale, then spending weekends and evenings hosting events for potential donors, businessmen and society aspirants who might give money to the Isolation Block. Colebrook needed private funds for his work.

All of his efforts kept the rate of childbed fever low at Queen Charlotte's. But he could not get rid of it entirely. Despite everything he did, childbed fever became established and remained intermittently epidemic at Queen Charlotte's Hospital, as it was at virtually every other maternity hospital in England, in Europe, and around the world.

In the early 1930s, every woman who entered a hospital to have a child was taking a risk. But still they kept coming to the hospital to have their babies, weighing the danger of childbed fever against the benefit of weeks of pain medications, nursing care, and weeks of bed rest, and deciding—just as the poor women in Paris had decided centuries before—to risk a "scientific," "modern" delivery.

CHAPTER TEN

AFTER FOUR YEARS they had found nothing. Domagk had worked steadily, Klarer furiously, the chemist synthesizing more than five hundred new substances, the physician testing each one meticulously against a half dozen of the world's most dangerous bacteria, infecting, injecting, scrutinizing, and ultimately sacrificing thousands of test animals. Klarer was not the only chemist whose products Domagk tested. He did the same for new compounds provided by a number of other Bayer researchers. By the end of 1931, he had tested nearly three thousand chemicals altogether, and the results were always the same: The test animals died from bacterial infections at the same rate as his untreated control animals. The chemicals did nothing. True, there had been a few tantalizing leads, small and scattered effects; a few animals had lived longer. He had discovered an antiseptic wash that Bayer took to market under the trade name Zephirol. But his main goal—finding a substance that would fight bacteria inside the body—seemed no closer in 1931 than it had in 1927. Nothing he tested in animals was strong enough or safe enough to use in humans. Domagk's laboratory notebooks tell a story of endless disappointment. Thousands of pages of notes list at the top of the page the chemical being tested, date received, its source (all Bayer chemists had their own abbreviations tied to the internal nomenclature for the test chemicals: Klarer's new compounds were tagged with the initials "Kl," Mietzsch's with "M"), tracking number, and chemical structure, followed by a summary of Domagk's test results broken down by type of animal, type of bacterium, dose, mode of administration (by mouth, subcutaneous injection, or intravenous injection), results, and then, at the bottom of the page, a separate set of results for tests done in test tubes (in vitro) rather than animals (in vivo). Both types of test were important. The in vitro tests, basically mixing dilutions of the test chemical with live bacteria in a tube and seeing if it killed the germs, showed if the chemical had antiseptic properties. The in vivo tests showed if it worked in a living animal. These were two very different things. As Sir Almroth Wright had shown at Boulogne, a good external antiseptic did not make for a good internal medicine. The body was too complex, the tissues too diverse, the reactions too intricate for a simple in vitro test to mean much. On the other hand, it was logical to assume that any medicine that worked against bacteria in the body would likely also kill bacteria in a test tube. Comparing the two results might yield some sort of insight. Domagk was looking for any sort of deviation, any clue, any slight effect. In his lab notes, he used the German word *Wirkung* to designate "effect"; "without effect" was *ohne Wirkung.* Domagk usually abbreviated them as *W* and *oW.* Four years of records, 1927 to 1931, are filled with the abbreviation for failure: *oW. oW. oW.*

No one expected that Hörlein's grand strategy to find a breakthrough antibacterial drug would succeed quickly. Some thought it would probably take a decade to find a single salable drug. Some thought it would never happen. As 1931 turned into 1932, Domagk, Klarer, and the rest of the research team were not yet frantic, although they would have preferred some encouraging sign, some indication that they were on the right track.

When they had started their program in 1927, times had been different, the German economy starting to recover, inflation beaten, their nation returning to life. At the end of 1931, the world was gripped by the Depression. Money was drying up. Unemployment was rampant. Sales were down. Bayer and its parent, IG Farben, needed a profitable new drug.

Despite the repeated negative results, Domagk changed neither his methods nor his approach. He was a very methodical and very patient man. He was happy to have a good job during hard times. And he was doing exactly what he had wanted to do ever since the war: seeking a way to defeat infection. He still viewed the process as a balancing act between invader (the germs) and host (the patient or test animal), with an antibacterial medicine acting as a way to tip the balance. That underlined the importance of testing in live animals; only here could the give-and-take among host, germ, and drug play out properly. Domagk often skipped or abbreviated his in vitro tests; he never failed to run tests in animals.

One issue was the endless flood of new chemicals. As they searched for improved substances for agriculture, industry, and consumer goods, the chemists at IG Farben and elsewhere were creating hundreds of novel molecules every month. They could not all be tested for medical effects. It would have taken a hundred pathologists like Domagk a century just to get through the backlog of chemicals that had already been created. So Domagk and Klarer picked and chose, read the literature, made educated guesses, attended conferences, and followed up leads uncovered by other researchers. They looked hard at any compound that any researcher reported as having antibacterial properties. They reexamined all of Ehrlich's work and pored over Roehl's old results. Following Ehrlich's lead, they often followed the trail of dyes that attached specifically to bacteria. Once they had a starting chemical, Klarer would alter its structure in ways he thought might improve it effectiveness, mixing it with other chemicals, reacting it with acids or bases, dissolving and condensing, heating and cooling, adding a carbon group here, removing a nitrate group there. If they got any sort of improved result, the altered molecule would become the new focus, new variations would be made around its basic structure, and the testing and refinement process would be repeated. In theory this would work them closer and closer to something that might be good enough to go into production. It was a matter of small steps, trying things until they found some small effect, incrementally improving it, and hoping for success, a chemical that could fight infection inside the body without being too toxic. Some side effects were to be expected, of course. It was too much to hope that a laboratory chemical could be given to humans without side effects.

One set of compounds they studied contained gold. Gold, like another heavy metal, mercury, had been considered a curative for thousands of years (some gold-containing compounds still are used today in the treatment of joint disease). Koch himself had investigated gold's ability to kill tuberculosis germs before discarding chemicals containing the metal because of unacceptable side effects, rashes, kidney problems, and an unpleasant condition with the evocative name "gold intoxication." Klarer and Domagk revisited the field, making and testing a number of new gold-containing compounds. While some worked in vitro, they all proved too toxic in animal tests. They turned to acridine dyes, a group that had started with Perkin's mauve, a family of dyes that Ehrlich and others had found to have an effect on parasites and bacteria. Some of Klarer's new acridines worked in the test tube; none stopped disease in mice. They tried a series of new quinine derivatives. Nothing. It was enough to make even the most hopeful give up hope. By 1930, as one American researcher noted, "[I]t was the almost universal opinion of physicians that nothing could be discovered which would be effective against the ordinary diseases produced by bacteria."

Domagk, however, remained committed to his search. He believed in the test system he had perfected, and he believed in Hörlein's vision for industrial research into cures. He had seemingly endless supplies of chemicals to test, animals to test on, and—as long as Bayer believed in him—money, support, and time.

Perhaps the most important factor was Hörlein's optimism. Hörlein, master of the research enterprise at Elberfeld, had set up the system, and Hörlein kept the money flowing to fund it through years of frustration. In his case the optimism was buttressed by proof that his system worked. No breakthrough medicine to use against bacteria had yet been found, but Roehl's old tropical-disease unit continued to uncover profitable antiparasitic medicines. Fritz Mietzsch had just synthesized a strikingly improved antimalarial, not only more powerful than Plasmoquine but easier to use; it killed the malaria parasite during the point in its life cycle when it was actually causing the main symptoms of the disease, making the timing of the medicine far easier for physicians. Atebrin, as Bayer named it, was released in 1930. It was not a perfect drug: It was expensive, for one thing, and dosing had to be close to toxic levels for the medicine to work well. It could cause diarrhea and vomiting, and it could turn skin yellow. In the army it was rumored to cause impotence, so German soldiers sometimes refused to take it; in military camps in malarial foreign areas, Atebrin often ended up in the toilet rather than in the patient. But when it was taken properly, it worked. It was the best malaria medicine on the market, a terrific sales success for Bayer, and it remained a very profitable drug through World War II.

If his new-drug research system worked for parasitic diseases, Hörlein believed, it would work eventually for bacterial diseases. False starts and dead ends were part of the game. He hewed to his vision, kept the upper administration happy (helped by occasional successes like Atebrin), and encouraged his team. The rewards would come. So his chemists kept pumping out new compounds, the animal staff kept breeding thousands of new mice and rabbits, and Domagk's operation continued to infect them and try to save them. The goal was simple: a new drug that could generate profits for the firm. Hörlein's system was going to pay off. He knew it. It was already working.

It just wasn't working for Domagk.

THE TROPICAL-DISEASE group was housed next door to Domagk's new laboratory, and a certain amount of communication took place between them. All of Klarer's chemicals, for instance, were provided not only to Domagk but also to the tropical-disease unit for testing against parasites. Although Mietzsch worked primarily for the tropical-diseases group, many of his chemicals were also tested against bacteria by Domagk. Mietzsch and Klarer worked at neighboring benches up in Workroom 4, and soon they began sharing techniques and approaches.

Domagk also looked to the tropical-disease group for starting points. Roehl had worked through a number of acridine dyes before Domagk and also had looked at another family of dyes, called azo dyes, before his death. Roehl had become interested in azos through his old boss Ehrlich, who found that one of them, called trypan red, had a notable effect against sleeping sickness in mice. Unfortunately, the effect was far weaker in humans. Other researchers had reported minor antibacterial effects of azo dyes as well. One form called chrysoidin had excited some interest. It was especially toxic to bacteria in the test tube, although it also had significantly toxic side effects in animals.

Late in 1930, after the gold and quinine tests had been abandoned, Klarer began working with azo dyes. These chemicals had a number of advantages. They were less toxic than many of the compounds the team had been exploring. Chemical variations were relatively easy to make. The core of the molecule—two carbon rings linked by double-bonded nitrogen atoms (this double bond between the nitrogens, the azo link, gave the family of dyes its name)—was like the frame of a bicycle. A chemist as talented as Klarer could easily change the wheels and gears, customize the handlebars and seat, add a cart in back or a basket in front, make a thousand variations on the core structure. True, azo dyes were a bit difficult to purify, and they had a bad habit of staining test animals various colors. Those were minor setbacks. More important, there were small hints from the beginning that azo dyes could kill bacteria. They were not powerful enough yet, but they might be made stronger. As Mietzsch put it, the chemist's goal was to put "the right chemical substituents in the right position on the azo group, when it is a question of bringing out the slumbering chemotherapeutic characteristics." At Bayer they believed that Panacea lived, perhaps within the family of azo dyes. It was just a matter of waking her. A steady stream of new azo dyes began flowing in 1931 from Klarer's workroom to Domagk's testing labs.

From the start the results were tantalizing enough—azo dyes sometimes killing bacteria in test tubes, other times having a weak effect in animals—to put Klarer into one of his frenzied states, a rapture of synthesis. Most research chemists did well if they created and purified one or two new molecules a week. Klarer could do that in a day. In 1931, hot on the trail of azo dyes, he began working to his physical limit, creating not only a new azo-dye derivative every few days—sixty-six of them over an eight-month period—but at the same time, because azo dyes were only one of several trails he was following, pumping out twice that many non-azo chemicals. It was a virtuoso performance. Some of his azo dyes began showing activity against tropical diseases, fighting bird malaria, rat leprosy, sleeping sickness. The results with bacteria continued to be spotty.

In the summer of 1931, something bigger began to emerge. When test results came back for the 487th chemical Klarer had synthesized at Bayer (Kl-487 in Domagk's notebook)—somewhere around the 100th azo-dye derivative the chemist had made, this one with an atom of chlorine attached—there was finally reason for Domagk to put a *W* in his notes. *Wirkung*. A definite effect. It wasn't as strong as they needed—Kl-487 worked against only one kind of bacteria, Domagk's superstrep, and it worked only in high doses—but something was there. The results were tallied on August 4, when most lab personnel were on summer vacation. When Klarer returned, he attacked the azo dyes with renewed zeal, trimming a side chain here, bonding a new one there, rebuilding and restyling Kl-487. He was doing nothing but azo dyes now. In the first three weeks of September 1931, he delivered fifteen new azos, producing every working day one new substance never before seen on earth. When Domagk tested them, he found the results perplexing. Some of Klarer's new molecules had a slight effect on bacteria, but others, seemingly close in structure, none at all; some stopped strep, others affected other bacteria; some killed bacteria in the test tube, others did not. There seemed to be no rhyme or reason. Klarer kept at it, but despite his best efforts, he could not seem to make the effects grow stronger.

On the eighteenth of September, Klarer delivered Kl-517 for testing. Like Kl-487, it included a chlorine atom; the chemical difference between the two molecules was slight. But the slight variation somehow made a significant difference. When Domagk saw the animal results in October, he was elated. He wrote "*Strepto!*" in his lab notebook, followed by several test results marked with a "*W!*" Kl-517 had protected some mice from a streptococcal infection—afforded not just a longer life, but complete protection. The chemical worked at lower concentrations than others, and it worked even when the substance was given by mouth (the ability to give a drug by mouth was an important sales advantage). Kl-517 again was not perfect—it worked only on his strain of test strep, was still not strong enough for the marketplace—but now Domagk was certain they were moving in the right direction. It was followed by more encouraging results, especially Kl-529, in which Klarer attached two chlorine atoms instead of one, which seemed to extend the effectiveness beyond strep to other bacteria. Klarer was again working at top pace. He was writing progress reports on so many azo dyes that he had a little rubber stamp made with the core azo structure; this he inked and stamped into his reports, then added the side chains by hand. On the eighth of November, 1931, Bayer filed a patent for a process to make the new chemical. As usual at Bayer, it was filed under the names of the chemists who synthesized it, in this case both Klarer and Mietzsch. Domagk's name did not appear.

Then, for some reason, the positive results stopped.

None of Klarer's newest variations were working. After a few weeks, Domagk went back and retested Kl-529. This time his results were not as encouraging as they had been in the first round. In December he retested again. This time Kl-529 did not seem to work at all. This was precisely what was *not* supposed to happen in Domagk's carefully refined test system, which he had devised specifically to avoid any sort of random variation. His lab notebooks now began to show uncharacteristic question marks along with strings of *oWs*—*ohne Wirkung,* without effect.

No matter what they did, Klarer and Domagk could not seem to get the azo-dye effort back on track. By the end of the year, the whole series began to look like another dead end. Not only were they unable to move toward more powerful variations, Domagk could not even replicate his earlier results. By the time they rang in the New Year of 1932, the trail had gone completely cold.

There had been an effect—not a real breakthrough, but at least something—and they had lost it. Klarer, in his year-end research report to the administration for 1931, did not even bother to mention his work on the azo dyes, but something kept Domagk from giving up. On one level he wanted his test system to work perfectly, and the azo results were perplexing enough to make him return simply to see what had happened. Perhaps he saw something in the results, something no one else could, the way he found bouquets of four-leaf clovers where others walked by. Whatever the reason, through early 1932 Klarer kept making azo dyes, and Domagk kept testing them.

The results remained confusingly erratic: an occasional weak effect on some kinds of bacteria, an occasional *W*, but generally no effect at all. In April, Klarer tried replacing the chlorine with an arsenic atom attached to the azo core, creating Kl-642, a dye that once again seemed to work a bit. Domagk found that it was effective against many kinds of bacteria but, like all arsenicals, was also highly toxic, too poisonous to try on humans. Klarer dropped arsenic and tried moving chlorine to other positions. He tried attaching iodine to the azo core. He tried potassium. He tried varying the length of side chains of carbon, capping them often with a nitrogen group, which seemed to help provide a weak effect. He tried shortening the side chains. He tried moving them around. He tried to find a pattern. None appeared. By the fall of 1932, somewhere around Kl-700, even Klarer was running out of ideas.

Then Hörlein had Klarer in for a chat. Perhaps the older man sensed that his talented young chemist—so much like a racehorse, so fast, so high-strung—was becoming discouraged. Perhaps, given that he himself had started his career at Bayer trying to make azo dyes more colorfast, he took a special interest in the azo work. It could have been that Klarer had asked for the meeting to complain about the progress he and Domagk were not making. At Bayer, after all, a significant portion of any chemist's income came from a share of money from profitable compounds they patented (all of Bayer's patents were taken out in the name of the discovering chemists to ensure that they shared in the profits). A percentage of the income from a fast-selling medicine like Atebrin could make a chemist like Mietzsch quite comfortable for years. Klarer, whose work with Domagk had yielded nothing of value, was scraping by on a minimal salary. Whatever the reason, Hörlein and Klarer spoke about the azo dyes. Klarer said later that it was Hörlein who brought up the idea of sulfur. Back in the old days, Hörlein and other chemists had made azo wool dyes more resistant to fading by attaching side chains containing sulfur. They had patented a few. If these sulfur-containing azos stuck more firmly to wool, perhaps they might stick more firmly to bacteria, providing a more reliable medical effect.

In the first week of October 1932, Klarer began attaching sulfur-containing side chains to his azo dyes. One of the first he made was an azo created by attaching para-amino-benzene-sulfonamide, a molecule more commonly called called sulfanilamide, one of the compounds Hörlein and the other chemists had used twenty years earlier. The sulfanilamide molecule itself was nothing special. In fact, it had been common around dye factories ever since a Viennese chemist named Gelmo first made and patented it back in 1909. It was long since out of patent and available in bulk quantities. Sulfanilamide—often referred to as sulfa—was relatively easy to make and cheap to use. It linked easily to other molecules. There was no problem integrating it into as many azo dyes as Klarer wanted.

The first sulfa-containing azo dye to reach Domagk in early October 1932 was Kl-695. Recollections differ as to what happened next. It appears that Domagk took an autumn holiday around the same time that the batch of a half dozen new chemicals, including Kl-695, came into the lab. While he was gone, his assistants kept up a full schedule of animal tests. Margarete Gerresheim, one of Domagk's top lab workers, remembered testing one of Klarer's new azo drugs—possibly Kl-695—on mice. The bacterial-disease panel was varied a bit from test to test. Sometimes Domagk's superstrep was included, sometimes not. Lately Domagk had not been doing strep tests on most of Klarer's new chemicals. Luckily, the tests *were* done on this batch. When the results came back, Gerresheim and the other assistants saw one very striking exception to the usual cages full of dead mice: the strep-infected animals who had received Kl-695. This one test batch not only survived but were, she told an interviewer many years later, "jumping up and down very lively." When Domagk returned from his vacation, Gerresheim proudly presented him with a large table summarizing the latest results. She told him, "From now on, you will be famous!"

Domagk's lab-notebook page for Kl-695 is different from the thousands of others that preceded it. He was a devil for consistency, always noting the results of tests on his bacterial panel—with its usual depressing list of *oW*s—down the left side of every page. On the page for Kl-695, however, most of the bacterial-panel results are pushed over to the right. In their place on the left is a long list of results for one kind of infection only: strep. And all down the list are *W*s. *W*s with plus signs. *W*s with double-plus signs. *W*s with triple-plus signs. Kl-695 worked like nothing Domagk had ever seen—like nothing anyone had ever seen. The chemical protected mice completely from strep infection. It protected them when delivered by syringe. It protected them when delivered by mouth. It protected them at every dose level he tried. The mice were not only alive but in perfect health. There were no apparent side effects. The results were so perfect it looked more like a mistake than a real experimental finding. Perhaps something had gone wrong in procedure; perhaps the test group had not received the proper dose of bacteria. Domagk immediately ordered retests with a wider range of Kl-695 concentrations. Again the drug kept the test mice alive, even in doses smaller than before. Again there was no apparent toxicity. Strangely, it did not kill strep in a test tube, only in living animals. And it worked only on strep, none of the other disease-causing bacteria. But, given the number and deadliness of strep diseases, it worked where it counted. It kept working through three series of retests. Hörlein was informed; when he heard the good news, he asked Domagk to keep quiet about his findings until they knew more.

Domagk then went back and ran strep tests on every chemical Klarer had given him for the past month or more, azo or not, plus everything every other chemist had sent him. All the animals died. The superstrep test was working perfectly. But Kl-695 continued to protect the mice.

That was standard operating procedure at the firm. Domagk was not much worried about patents. The important thing for Domagk was that nothing had worked for five years and now suddenly everything seemed to be working. Klarer now made variations on Kl-695, finding that as long as sulfa was attached to the azo-dye frame in the correct position, the drug worked against strep. Attaching sulfa to an azo dye—any azo dye—somehow transformed it from an erratic, ineffective chemical into an efficient anti-strep medication. The sulfur-containing side chain simply had to be attached in one particular spot at one end of the azo dye; as long as it was hooked there, it worked—against strep. Klarer kept exploring. Perhaps tweaking the molecule a different way would make it effective against more kinds of bacteria. Perhaps a variation might work against cancer. The guiding principle for both Klarer and Domagk was that the frame in the middle, the azo-dye core, was the power center for the medicine. The side chains, the add-ons like sulfa, were keys to turn it on. More needed to be known about how this sulfa key unlocked the power of the azo dyes.

Klarer made seven more anti-strep drugs using Kl-695 as a starting point. Some were even stronger than Kl-695. One in particular, Kl-730, was the most effective and most consistent anti-strep medicine anyone had ever seen. But Kl-730, like all the others, was selective for strep only, ineffective against cancer, tuberculosis, staph, pneumonia, and all of the other bacteria in the standard test panel. It was confusing. Staph and pneumonia in particular were germs that appeared to be built like strep and reacted to most dyes like strep. Still, the new medicines did not work well against them. And it continued to be bothersome that the new sulfa compounds were harmless to strep in vitro. The new chemicals appeared to have no antiseptic qualities whatsoever. Inside the body they were powerful medicines; outside they did nothing. It made no sense.

By late November all attention was focused on Kl-730. It was a dark red azo dye, almost insoluble in water, tolerated by mice in very high doses without significant side effects. Domagk remembered the mixed elation and disbelief of these early experiments, writing that when he and his assistants first saw the cages of healthy mice, "We stood there astounded, as if we had suffered an electric shock." Domagk's notebook page for Kl-730 is packed with scribbles in different kinds of pencil, ink, and what looks like red crayon. Kl-730 results from various dates are sandwiched together, overlapped and overwritten. He remembered examining the animals in test after test, opening them and peering inside, taking samples to the microscope. "We dissected until we could no longer stand on our feet and looked through microscopes until we could no longer see." The control mice who received strep but no drug were teeming with bacteria. Those that received Kl-730 had healthy organs, healthy tissues—and no sign of a single living streptococcus. In these animals Domagk saw a few bits and pieces that might once have been bacterial cell walls, detritus still being attacked by the animals' immune systems, but that was all. Looking inside these mice was like looking at a battlefield after the battle. The bacteria had been utterly defeated.

Later journalists made up their own version of the story, turning it into a Christmas tale that hinged on a single, pure moment of discovery. In the most widely repeated version of the story as it appeared in newspapers and magazines, readers see Domagk, just before Christmas 1932, handpicking twenty-six white mice from his animal facility and personally administering doses of Kl-730, threading a small rubber tube down the throats of twelve of the animals, and delivering the drug directly into their stomachs, the other fourteen controls given no medicine. Domagk then shaves a tiny patch off each animal's stomach and injects a fatal jolt of strep—enough bacteria to kill each mouse ten times over—into the lining of its abdominal cavity. Within a few days, as Christmas approaches, all the unprotected mice are dead. But without exception every one of the dozen mice treated with Kl-730 are alive a full week later. There is joy in Elberfeld.

It is true that as the Christmas holidays approached in 1932, Bayer was abuzz about the first drug to ever work against bacteria. After Domagk showed the Kl-730 results to Hörlein, a series of meetings was set up with Bayer patent attorneys, marketing men, and higher administrators. Domagk was invited to present his findings. A preliminary name for a new drug, Streptozon, was quickly approved. Klarer began working with Mietzsch to keep up with the demand for more variations on the Kl-730 structure. By mid-December they had come up with a dozen new versions. It became clear that the Bayer chemists could make effective antistreptococcal medicines whenever they wanted to by attaching sulfa to an azo dye in the right place. You could vary almost anything else and still have something that worked.

This was fine, but it also presented the company with a dilemma. Once the new medicines were made public—as they would be as soon as the patent application was approved and entered the public record—it appeared that anyone with some basic skill in organic chemistry would be able to make his own versions of sulfa drugs. Bayer could not patent every one of them.

They could, however, patent the best. So while the business office thought about the problem, Klarer and Mietzsch turned out sulfa-containing azo dyes as fast as they could. Domagk was given the task of determining which one was most effective. Rather than the single determinative experiment touted in the press, the Christmas holidays in Domagk's lab in Elberfeld were celebrated with a mass of concurrent strep tests on Kl-730, Kl-695, and Kl-713, -721, -722, -724, -725, -726, -727, -731, -732, -733, -737, -738, -745, and -746. Within a period of two weeks at the end of the year, hundreds of mice were infected, sacrificed, and scrutinized. Results were checked, cross-checked, and compared.

Domagk reached his conclusion on December 27. Kl-730 was the best. Hörlein appears to have anticipated the decision; to speed the process, perhaps to underline its importance, he had somehow arranged for a patent to be filed for Kl-730, again under the names of Klarer and Mietzsch, on Christmas Day 1932.

II

The Right Side

A discovery is said to be an accident meeting a prepared mind.

Albert von Szent-Györgyi

CHAPTER ELEVEN

GERMAN CHEMICAL PATENTS were often small masterpieces of mumbo jumbo. It was a market necessity. Patents in Germany were issued to protect processes used to make a new chemical, not, as in America, the new chemical itself; German law protected the means, not the end. Bayer, in its German patent application, had to describe the laboratory procedures used to make Kl-730 clearly enough to establish its rights, but not too clearly. If the process were outlined in too much step-by-step detail, anyone could make the new chemical, study its structure, find an alternative way to manufacture it, and patent the new process—shrinking the value of the original patent. In Germany, therefore, chemical patents were often written in ways that described the process while at the same time lessened the chance of duplicating it. The wordsmiths in Bayer's patent office were masters of twisting, subtle language that rivaled that of the most obscure modern novelists.

But then most things about Kl-730 were confusing. It was expected that if it stopped one type of bacterium, it would stop others. It did not (at least not anywhere near as effectively as it did strep), giving weak, variable, generally negative results on other types of germs. It was expected to have side effects. It did not. It was expected that if it killed bacteria in a living animal, it would kill them in a test tube. It did not. No one knew how it worked, why it worked, or if it would work in humans.

There was much the company did not know, but Bayer pushed its new medicine toward the marketplace regardless. It was now out of the laboratory and in the hands of company executives charged with readying it for sales. Its new trade name, Streptozon—a play on its strep-killing focus—replaced Kl-730 in all company correspondence. Domagk and his entire group were able to take a deep breath and relax after the patent was filed. There were still plenty of chemicals to test, of course. Klarer and Mietzsch through the first months of 1933 continued making new variations, trying to figure out how adding sulfa gave the molecule its power, altering and retesting the Kl-730 molecule. They proceeded from the same paradigm that had guided them since the beginning of the azo work two years before: that the germ-killing power of the drug resided in the dye. The dye-as-medicine paradigm went all the way back to Ehrlich. This was a powerful idea within Bayer, a company built on dye research. It was obviously the dye that found and specifically attached to the bacteria. Klarer had created a number of azo dyes with limited germ-killing abilities. Without the dye, they all assumed, there could be no medicine.

So why was the sulfa important? While the marketing and sales staff prepared for a release of the new drug, the chemists and the pathologist kept tinkering, looking for answers, though with far less a sense of urgency. They assumed that sulfa was somehow a key that turned on the azo dye, that sulfa was a way to waken the dye's slumbering power. The chemists tried altering the sulfa part of the molecule and found that it was not the sulfur atom alone that mattered but the entire sulfanilamide structure (in which an atom of sulfur was attached in a specific way to oxygens, nitrogens, hydrogens, and a carbon ring); sulfur in other combinations did nothing. They confirmed that not just the structure but the positioning of the sulfa on the azo-dye frame was vital. If they moved it even a little, altered its relationship to the azo core by even an atom or two, the dye went to sleep again.

Once they knew the magic spot where sulfa worked, the chemists tried attaching other things there as well. And they began seeing some interesting results. A chlorine atom placed in that position offered some limited activity against staphylococcus. An iodine atom yielded a dye that showed some activity against tuberculosis. TB was a huge health problem, and Domagk was intent on solving it. The Bayer team's work with iodine-linked azo dyes would grow into its own track of investigation that would bear fruit eventually. But for now they were coming up with a number of weakly positive results. Nothing to add to Streptozon for the market—not yet, at least—but it looked as if Hörlein's gamble of pouring money into this search for antibacterials was finally going to pay off. It looked as if Streptozon was simply the first of what could be a series of azo-based antibacterials effective against all sorts of diseases. It looked as if Bayer had prospected its way into what could be an incredibly rich pharmaceutical field.

Continued experimental attention was given to Streptozon itself, extending and confirming the initial findings, varying the doses to find a lower limit for the drug's effectiveness, adding rabbits to the animals tested. More complete toxicological and pharmacological studies were performed to see exactly what the drug did in (and to) the test animals. The findings reaffirmed that the drug was amazingly powerful against strep and astonishingly safe for animals. The company could not have wished for a better medicine. Damage was limited to some minor effect on the kidneys at very high doses. Streptozon survived the acids in the stomach, entered the bloodstream easily, traveled quickly through the body, and was excreted in the urine. Fast excretion was a good thing: The drug would not build up in tissues. The worst side effect was more cosmetic than toxic: The dye turned the test animals' skin pinkish red for a few days. Streptozon was incredibly good. Almost too good to be true.

Still they kept looking for azo dyes that would stop not only strep but other common pathogens as well. While they calmly and deliberately continued the search, the rest of the company geared up for Streptozon's introduction to the market and the promise of enormous, colossal, unheard-of profits.

ON FEBRUARY 20, 1933, a few weeks after Adolf Hitler was named chancellor of Germany, a few days before the Reichstag fire, Carl Duisberg, the great industrialist, the head of all of Bayer, proudly threw the switch on the biggest electric sign in the world: twenty-two hundred lightbulbs suspended between two smokestacks at his huge Leverkusen factory. It was a circle twenty stories tall containing the famous Bayer cross—two BAYERS, one vertical, one horizontal, crossing at the Y—a flashing symbol that looked like a giant Aspirin tablet. First the circle blinked on, then the BAYERS blinked on, then both off, the pattern repeating all night long. It was the company's international trademark writ large, seen for miles reflected in the Rhine. It was Duisberg's notice to the world that the Bayer Corporation—not IG Farben but his Bayer, the company he had made into a giant, a survivor of war, economic upheaval, and corporate merger—was still alive and very healthy. With his company's symbol flashing behind him, Duisberg told the crowd of workers gathered for the occasion, "As the Southern Cross gives direction and hope to the mariner, may this 'Western Cross' in the heart of German industry shine out to the German merchant, the German entrepreneur, and the German workingman as a symbol of our courage and our confidence. May it also be a symbol for the rest of the world of the conscientiousness and quality of German work."

The quality of German work on Streptozon led next to yet another puzzle. In their continuing attempts to discover how the sulfa key switched on the dye, Klarer and Mietzsch kept clipping and fashioning the sulfanilamide molecule before incorporating it. At the end of March 1933, as part of this inquiry, they delivered two stripped-down chemicals to Domagk. These were not azo dyes at all, but sulfanilamide attached to other compounds. The first, Kl-820, did nothing. That was what they expected—without an azo dye, there should be no medicine.

But, amazingly, the second molecule, Kl-821—sulfanilamide linked not to an azo frame but instead to a relatively simple carbon-and-nitrogen string—worked, and worked extremely well. It stopped strep infections in both mice and rabbits. When the first test results came in in mid-April, Domagk immediately ordered retests in strep-infected rabbits and got even better results. The initial mouse results were noted on April 11, 1933, with an underlined *W* for the strep-infected animals. The initial rabbit results were tallied on April 12, a strep retest on April 26, a staph test on May 3. Domagk highlighted the second round of rabbit tests with a double-underlined *W.* Kl-821 worked at all sorts of dilutions. Kl-821 was a very powerful antibacterial —without an azo dye. The power of the medicine, it would seem from this molecule, was not necessarily in the dye. Which meant logically that the power was in the sulfanilamide.

This discovery should have shaken the Bayer drug-research program to its core, but for some reason it seemed not to have. What happened instead is that everyone seemed to pause for a moment, as if the entire company took a deep breath. Then Mietzsch and Klarer dove back into a frenzy of work, their efforts once again directed at azo dyes. The relatively relaxed pace of the past few months was over. The two chemists began making sulfa-containing azo dyes at a rate almost equal that of the previous fall. In early May there was another mass comparison of the latest variations. After summer vacation the chemists' pace accelerated even more, Klarer and Mietzsch concocting a new chemical every working day, all of them azo dyes, half of those containing some form of sulfanilamide, the rest with chlorine or iodine and then, once again, arsenic.

Why did they ignore Kl-821 and its obvious—at least in hindsight—implications? There is a chance that the results with that single chemical were seen as an anomaly and discounted. A few unusual, unreplicatable results are scattered throughout Domagk's lab notebooks, as they are through the lab notebooks of most scientists, occasional successes that did not hold up, occasional failures that could not be explained. Lab research is messier than most people think, and perfect results are rare. In all the excitement over the azo dyes, the results with Kl-821 might have been set aside and forgotten as another one of those oddities.

Perhaps it was a case of seeing what one wants to see. A sort of dye mystique ran deep in German chemistry. Dyes had made German organic chemistry profitable. Ehrlich had believed that dyes could be medicines, and a generation of researchers had followed his lead. Dyes had marked the trails that Domagk and Klarer had explored for the past five years. Even scientists sometimes see what they expect to see and discount things that do not fit their worldview. There might have been a strong inclination for the Bayer team to stick with what they knew.

But neither of these factors seems sufficient to explain Domagk's ignoring the Kl-821 results. He was too keen-eyed for that, too meticulous, too attuned to the slightest variation in test data. He was interested enough to retest it, then retest again, the results noted in pencil, black pen, blue crayon, some results underlined in red. It is difficult to believe that he simply ignored what he was seeing.

It is possible that the Bayer team *did* realize that the results they were seeing came from the sulfa side chain rather than the azo core and realized as well what this meant: that the world's first breakthrough antibacterial medicine was a cheap, unpatentable bulk chemical. Anyone could make sulfa. There was no way for Bayer to make a fortune off it.

Whatever the reasoning, the Bayer team went back to azo dyes. Klarer and Mietzsch spent the last seven months of 1933 producing seventy-six more new azos, many of them containing sulfanilamide. There is no public record of any further attempt to test sulfanilamide without its incorporation into an azo dye. No further non-azo sulfa drugs show up in Domagk's notebooks.

The pace continued into 1934, the chemists pumping out new azo dyes, Domagk testing them, retesting Streptozon, and then retesting Streptozon again. He was perfecting a new animal model of strep disease now, a strep-caused joint inflammation in rabbits, and he found that this could be cured with Streptozon as well. It was gratifying to see the results: Animals that were wasting away, their joints pus-filled and grotesquely enlarged, became completely healthy after only a few days of treatment. Again there were no serious side effects even with very high doses, no damage to organs, no loss of weight, heart and blood vessels fine, blood and urine normal, blood pressure unchanged. Streptozon was shown once again to be "an extraordinarily indifferent compound," as the toxicologists put it—at least in mice and rabbits. Humans might be another thing.

There was one important drawback, however. Streptozon was a solid that was easy to powder and form into brick-red tablets but almost impossible to dissolve in a liquid. Severely ill patients—those who were unconscious or delirious, those with swollen throats—were unable to swallow pills. Bayer needed a liquid form of the drug. It took more than a year, but Mietzsch finally came up with one in June 1934, a beautiful port-red fluid that could be injected via syringe. They called it Streptozon solubile. In tests it was just as active as the powder.

There was more good news. Bayer slowly, conservatively began letting Streptozon out for tests in human patients. Today the testing of new medicines in humans—clinical testing, as it is called—is a highly organized process with established ethics, regulations guiding the informed consent of patients, and detailed safety guidelines. In 1933, it was a far more haphazard business. It was still common for physicians or chemists who found a new medicine to test it first on themselves, on colleagues, even family members. As drugs became more powerful, so did side effects—and so did concerns about trying new chemicals on humans before they were proved safe. Bayer and other European chemical firms often went to Africa to do large-scale human tests. In Britain they used soldiers. In the United States, tests were done on prisoners and inmates in mental institutions.

None of these things were done with Streptozon. Instead the drug was leaked to hospitals quietly, on a case-by-case basis. Domagk talked about it to physicians he knew in the area around the Elberfeld plant, who in turn told others. Within a few weeks of the Christmas patent, word began to spread about a new drug that might cure strep diseases. Domagk started receiving letters from physicians begging to try it in patients dying of strep and other diseases. He personally distributed samples to a few nearby hospitals.

The first tests of Streptozon in humans were more humanitarian gestures than carefully run experiments. A young Düsseldorf physician named Förster in early 1933 was about to lose a ten-year-old boy to advanced blood poisoning when he heard about Streptozon from an older physician in his hospital, who had heard about it from his friend Heinrich Hörlein. It was not a strep disease—it was staph blood poisoning—but it was a hopeless case, the boy would likely be dead within a few days, Streptozon had some limited effect against staph, so why not? No one knew what a proper dose should be. Förster treated the boy like a very large mouse, extrapolating an effective dose from what had worked in a certain-weight animal up to the weight of the child. He gave the boy a red tablet to swallow, then crushed another half tablet in water, the particles refusing to dissolve but small enough to swallow easily, and had him gulp that. No one knew what to expect. Within a few hours, the boy's skin turned bright red. Förster gave him another half tablet. The next day the boy's temperature started to come down. After three days and a total of just four tablets of Streptozon, the infection, to everyone's surprise, seemed to have disappeared. The red-stained skin faded. The boy appeared to be totally cured. Förster reported the case to a meeting of dermatologists in May 1933—the first public announcement of Streptozon's effects in a human—setting off another round of requests for samples.

Domagk provided Streptozon to one of his friends, Philipp Klee, the head of internal medicine in the nearest large hospital, in the newly created town of Wuppertal (an amalgamation of several older villages, including Elberfeld). Domagk and Klee had similar temperaments, both reserved, sensitive, careful, and intellectual. They even looked alike, with high foreheads, upright bearing, receding hairlines, and piercing blue eyes. They shared an interest in art as well. Domagk had long appreciated modern painting, and Klee had married a Budapest artist, Flora Palyi. It did not matter much to either of them that she was Jewish.

Klee tried Domagk's new medicine on an eighteen-year-old girl who had entered his hospital with a severe strep throat infection. The bacteria were invading her body, creating abscesses behind her tonsils, threatening to get into her blood. Her temperature shot up, and her white blood cell count skyrocketed. The usual therapy, opening and draining the abscesses, helped for a while, but the girl relapsed. Her temperature rose again; a blood clot formed in her jugular vein (one of the ways the body tries to stop a blood infection); she was shivering and drenched in sweat. Then her kidneys started to shut down, and she stopped urinating. She was about to die. With nothing to lose, Klee dosed her with Streptozon. The next day her temperature fell to normal. Her kidneys restarted, and she began urinating "copious amounts," Klee reported. She was kept under observation and given continuing doses of Streptozon. A few weeks later, they could find no trace of strep in her system, and she was discharged. Klee got more of the drug from Domagk and began using Streptozon on patients suffering from a variety of infections, inflamed tonsils, boils and abscesses, fevers following septic abortions and childbirth. Not all of his attempts resulted in a cure. But most of them did. "The results were indeed miraculous," a medical historian wrote later. "To the properly skeptical and meticulous Philipp Klee, who had seen so very many of his patients die from exactly the same infections, this was the most wonderful experience of his life."

The word spread, and the requests for Streptozon grew to a steady stream. In Düsseldorf a senior physician in the skin department named Schreus—the same friend of Hörlein's who had gotten the drug to Förster—tried it against two of the worst infectious diseases in hospitals, erysipelas and cellulitis, as well as strep-caused joint infections and rheumatic fever. Again there were cures in most cases. He did find that the new drug had little effect on strep infections of the heart and was unable to reverse the most advanced cases of infection in older patients, after the circulatory system had begun collapsing or if their fevers rose too high. Overall it worked best against strep, as expected, and sometimes, although less often, against *Staphylococcus* infections. In younger patients, especially, its effects were incredible. Side effects were limited to some nausea.

The cases were individual, the amounts of drug used variable, the results communicated, if at all, in small meetings or in relatively obscure German medical journals. As Hörlein wished, the news about Bayer's new miracle drug remained contained and muted while his team continued to work. By the summer of 1933, Streptozon had been shown to be both effective and safe in animal tests and in at least a few humans. By the summer of 1934, the drug had been tried on a growing number of patients and was available to a handful of German physicians for testing in both tablet form and as an injectable solution. But in late 1934, fully two years after the initial patent had been filed for Kl-730, there had still been no widespread release of the medicine, no articles from the Bayer team relating their success published in any scientific journals, no reports of the animal tests, no general public announcement from Bayer that the company had discovered the world's first antibacterial medicine, a safe, effective compound capable of saving hundreds of thousands, perhaps millions, of lives each year. Bayer continued releasing other drugs—an injectable liver extract, a new antiepilepsy medication, an improved successor to Atebrin—in addition to pushing forward research into synthetic rubber and synthetic fibers. But the company kept curiously quiet about Streptozon.

Hörlein said later, when asked about the long delay, that within Bayer there was a level of disbelief that had to be overcome; extraordinary care, he said, had been taken in testing and retesting before making announcements because, in effect, Streptozon was too good to be true. The impression he gave was that skeptics within the company needed to be convinced by repeated and confirmed proof of the drug's power, especially concerning safety, before they would risk the firm's reputation by announcing it to the world. Hörlein's excuse is plausible, but more than one observer has pointed out that in fact Bayer did not seem to put much effort into testing the medicine's safety through extensive human trials, instead letting the drug trickle out on a case-by-case basis, lessening the chances of definitive large-scale experiments in human subjects that would have convinced the skeptics.

Another reason for the delay might have been that Bayer was hoping for something better. Mietzsch and Klarer's continued tinkering with azo dyes hinted at new forms that might be effective against gonorrhea, staph, and tuberculosis. As long as their work remained secret, Bayer would maintain its position as the only place in the world doing research on sulfa-containing azo dyes. Publicizing Streptozon threatened to draw the attention of the rest of the world's drug firms, possible poachers on Bayer's preserve. There was no way to protect the area forever, because their work already showed that any number of azo-dye derivatives could be active as medicines; Bayer could not patent them all. But delaying publicity about Streptozon gave Bayer time to find and patent the best of them. It was all right for Domagk to pass the drug out to a few of his medical friends for emergencies, and it was necessary to get at least some clinical backing for the eventual marketing effort, but there was no hurry. As long as the patent they had filed back in 1932 was still under review pending final approval, it would not be made public. They had time.

They could not keep it secret forever, though. Physicians using the drug were talking too much. As far away as England, by 1934, a few physicians and chemists knew that, as one put it, "something was brewing in the Rhineland." On December 13, 1934, Bayer's patent for Kl-730 was granted and entered into public record. Anyone who wanted to know how to make Streptozon, at least in vague terms, could now look it up. Publication of the patent seemed to push Bayer into finally releasing the drug widely.

It now had a new name, thanks to the advertising and marketing offices at Bayer. The most recent clinical tests had shown that the drug sometimes worked against staph and gonnorhea infections, and a new name was needed that was not limited to strep. The name they chose, Prontosil, also hinted at the fact that it worked quickly.

Then, two years after the fact, Domagk finally began writing his first article on the discovery. It was a strange piece of work. In it Domagk described only one of his scores of animal experiments, a test his lab had done on Kl-730 and strep just before Christmas 1932. He did not mention anything about the other eight azo dyes tested at the same time against the same germ, all of which gave equally dramatic results. He did not note any of the results in rabbits or give detailed information about the medicine's effects against any bacteria other than strep. Still, because of its importance in announcing the scientific methods used to find the world's first widely effective antibacterial chemotherapy, it has become a classic of medical history. "It is currently the general opinion that only protozoal infections can be attacked by chemotherapeutic means," Domagk began, and went on to review the many failures in the search for ways to attack bacteria, the chemicals that had proved ineffective or too toxic, before running through his results from December. The paper, "A Contribution to the Chemotherapy of Bacterial Infection," appeared—in conjunction with papers from Klee and others that detailed a few clinical successes—in a leading German medical journal in February 1935.

It was then generally ignored.

Part of the problem was the limited scope of Domagk's paper—one experiment only—and the too-perfect results he got, every control animal dead, every test animal healthy, which raised the natural skepticism of physicians. The results came from an industrial lab, which in itself was suspect. There was no explanation given for the two-year-plus delay between the date of the experiment and the date of publication. There was no indication of how the drug might achieve its surprising effects. There was the puzzling fact that the medicine did not work in the test tube. As Domagk noted in his paper, "Whether Prontosil acts directly or indirectly against the pathogen in the body cannot be decided as yet. It is remarkable that *in vitro* it shows no noticeable effect against *Streptococci* or *Staphylococci*. It exerts a true chemotherapeutic effect only in the living animal." Then there was Domagk's tone, oddly restrained, mentioning the shortfalls of the drug in treating anything other than strep and calling for "close cooperation" between physicians and bacteriologists to identify the source of infection, a manner of writing described by one historian as "extremely conservative as to conclusions."

A single, suspiciously good animal experiment, a number of significant questions, and a handful of anecdotal reports of human cures were not the stuff of wild enthusiasm. No wonder the paper was greeted with what appeared to be a collective yawn.

It was not, however, totally without effect. The German journal in which it appeared was respected and widely read, so many people saw the paper even if few were enthusiastic about it. Its publication seemed to open the door for more clinical reports, which now began showing up regularly in medical journals: Dr. Gmelin in Essen cured blood poisoning in children with Prontosil; Dr. Veil in Jena reported positive results treating rheumatism; Dr. Anselm in a women's clinic, cures for patients with childbed fever. Physicians were getting good results in Dortmund, Frankfurt, Cologne, Göttingen, Munich, and Berlin. Veterinarians began reporting success in treating livestock. Bayer finally began rolling Prontosil out into the marketplace, relatively slowly at first, in Germany, then throughout Europe. Instead of requests for samples, Domagk now received correspondence from German physicians asking for advice on proper dosages and target diseases.

The grand dream of an effective antibacterial chemical—history's first—was about to be realized. Despite all the worries, skepticism, and disbelief, it appeared that Prontosil really cured human disease. A nontoxic internal disinfectant exquisitely targeted to bacteria, Ehrlich's long-sought *Zauberkugeln,* had finally been found. Panacea, after thousands of years of failed attempts, had finally awakened.

CHAPTER TWELVE

On October 3, 1935, Heinrich Hörlein delivered a lecture to the prestigious Royal Society of Medicine in London on the chemotherapy of infectious diseases caused by protozoa and bacteria. The talk was heard by many of the leading health experts in the United Kingdom. The news of more human cures in Germany was beginning to draw attention in other European nations, and Hörlein used his Royal Society appearance both to inform the curious and to rouse enthusiasm for the new drug Prontosil among practicing physicians. He reviewed Domagk's results, published in a German journal eight months earlier but still news to many in the audience, stressed that the new azo dyes were "remarkable for their non-toxicity," and made certain that the British physicians in the audience knew that the Germans had seen "remarkable effects" using Prontosil in a number of streptococcal infections, where "in many cases it has proved a life-saving agent." He noted as well that the drug was almost entirely strep-specific, having little effect on other bacterial infections—with the sole exception of a limited benefit against staph.

Sir Henry Hallet Dale, first director of England's National Institute for Medical Research, was impressed. He already knew more than most Englishmen about Prontosil. He had been introduced to the drug back when it was still called Streptozon, well before Domagk's first paper appeared. On a visit to Elberfeld in 1933 or 1934, he saw for himself the enormous effort German industry was putting into pharmaceutical research and heard about the dye-medicine that could cure strep infections in mice. Immediately after Domagk's article appeared in February 1935, Sir Henry asked for samples. He knew that the clinical results reported in Germany were inadequate—too few patients, no proper controls—and he wanted to get the drug to a British researcher for more thorough testing. The best man for studying a new drug for strep infections, he thought, would be Colebrook, leader of the investigations into childbed fever at Queen Charlotte's.

Sir Henry was a forward-looking man. He believed that medicine was finally about to take its place as a science as much as an art. This new, modern medicine, armed with the latest and most powerful tools from biology, chemistry and physics, X-ray machines, centrifuges, chromatography, immunological techniques, research into the molecules of life, controlled experimentation—the methods and approaches of the established "hard" sciences—was going to usher humanity into a golden age of health. These were the tools Sir Henry had used while researching nerve conduction, and they had worked. Now there was talk of a Nobel Prize for his research. Sir Henry's technological optimism was positively German, confirmed, perhaps, during the months he had spent as a young man in Ehrlich's lab, a wonderful introduction to dyes and medicine. Sir Henry had seen the future unfold when he saw Ehrlich's methylene blue bringing to light the fine pattern of nerve cells under a microscope. He still kept up with the latest news from Germany.

After reading Domagk's article, he quickly wrote Bayer. "Colebrook is interested to try a new preparation issued by the Bayer Products Company called 'Prontosil'; and I wrote to Dr. Hörlein, of Elberfeld, to obtain some," Sir Henry told a colleague. "He replies that they will be delighted to supply all that is required."

Delighted, perhaps, but not swift. The remainder of March passed, then April, May, the rest of the spring, and a good part of the summer of 1935 before Colebrook finally got the promised Prontosil tablets plus a supply of the deep red liquid that the Germans were still distributing under the old name "Streptozon solubile." Colebrook's coworker Ronnie Hare thought that the Germans seemed "less than willing" to let Colebrook have the Prontosil, a delay that Bayer's representative in England blamed on a "hitch in the supply." The reasons were not important. After his failure to save any new mothers in the Isolation Block with arsenicals, Colebrook was simply happy to begin testing the promising new drug. He, too, had seen Domagk's paper and the accompanying clinical reports as soon as they appeared—and like many readers was unimpressed. The human tests were too sketchy to be convincing. Domagk's own findings were both too much and too little: those suspiciously perfect results, yet in only a single mouse test; no references to similar basic research; no ideas on how this medicine picked strep to attack; no idea of how or why the drug worked. Prontosil looked like, as Hare put it, "another of those damned compounds from Germany with a trade name and of unknown composition that are no use anyway."

But Colebrook was willing to try. It was, after all, reportedly nontoxic, so why not? He got his first batch of Prontosil on July 18, 1935, and immediately started work on strep infections in mice, seeing if he could replicate Domagk's initial findings. What he saw made no sense when put up against the German reports. The new drug did not seem to cure strep infections at all. The mice he treated with Prontosil died almost as fast as those he did not. Three months later, after Hörlein's Royal Society speech, Colebrook was still struggling to get some positive results.

"I keep receiving requests from clinicians for permission to obtain supplies of these substances for use in patients. These requests have followed, of course, on the recent favorable communication about Prontosil to the Royal Society of Medicine," Colebrook wrote a colleague two weeks after Hörlein's October appearance. "I am sorry about delay in reporting but one hesitates to send in a completely unfavorable report without adequate evidence. Up to date my experiments with mice give no support whatever to the German claims. . . . I think you will do nothing but good by checking the optimism and hopefulness of enquirers meanwhile."

He did confirm Domagk's finding that Prontosil had no effect on strep in a test tube. But in Colebrook's hands the medicine had no effect in mice either. No matter how he infected them or how he varied the doses of dye he gave them, they died. He was using bacteria that he had gathered from his own patients, strep from childbed fever victims that had been cultured in his laboratory for some time. And that turned out to be the problem. In an attempt to resolve the discrepancy between his results and the Germans', he sought new strains of strep. When a fellow researcher provided him with a far more virulent strain, more like Domagk's superstrep, Colebrook's mouse results suddenly began mirroring the Germans' (although they were never quite as flawless).

This was somewhat more encouraging, but still not enough to move ahead to human tests. Colebrook would have preferred that the drug work on the types of strep he isolated from his own patients; with no indication that it could cure the particular strains attacking women in his care, what was the point of giving them the medicine? Colebrook decided to find out more about how living animals metabolized the drug, what it did to the blood, how it was excreted, before he tried it on his childbed fever victims. Pressure meanwhile was growing in the form of requests for the drug from British physicians who continued to hear stories of wonder cures in Germany. In early November, Colebrook received a letter from Sir Henry Dale, who told him about a London physician "pleading for help, on the ground that he has been suffering from a streptococcal infection, starting with an osteomyelitis of his jaw, a large part of which he has lost without, so far, ridding himself of the infection." After hearing that Hörlein had spoken of lifesaving cures with Prontosil, he, like others, began searching for the drug. Dale bemoaned Hörlein's "stupid statement" but also let Colebrook know that his colleague, the ailing physician, was begging to be made part of whatever tests Colebrook was running. "He is evidently desperate, and willing to have any experiment tried on him," Sir Henry wrote. Colebrook replied with his regrets. He was still not ready for tests on humans.

Finally, in January 1936, Colebrook reported the results of almost six months of mouse tests. "Prontosil certainly has an effect in mice infected with highly virulent *Streptococci* (three strains), although very much less than that reported by the Germans," he wrote Sir Henry. "It apparently has no effect, or as good as none, upon *Streptococci* of ordinary virulence as isolated direct from human infections. It seems to me, therefore, unlikely that we shall get any good results in human cases, but since its mode of action is quite obscure, there is a possibility, and I am going to try it, but I am afraid it will take several months."

He was being purposely low-key, trying to deflate overenthusiasm while at the same time beginning to see some very interesting results. The first test in a British subject was not on a childbed fever victim at all but came about—like so many things related to Prontosil—by accident. On January 6 his colleague Ronnie Hare accidentally pricked himself with a glass shard contaminated with some of the ultravirulent strep they were testing. Within a day or two, the infection had spread through his bloodstream, and it began to look doubtful whether he would live. In desperation Colebrook gave him the experimental drug. Hare was too weak to protest when Colebrook dosed him with Prontosil, both intravenously and orally. "I thereupon became a bright pink in color and felt so much worse," Hare remembered. But the nausea quickly passed, and his fever subsided. To everyone's surprise the blood infection, over the course of a few days, disappeared. Not only did he live, but he also retained the use of his hand—a result far better than any of the physicians around him had ever seen.

Colebrook, like all the early researchers, did not know exactly what doses to give his patients, how much they could tolerate, how much might be toxic. He did not know whether oral or intravenous Prontosil was better. He did not know how long it might take to have an effect, or how often to administer the drug, or how long to continue the treatment. He did not know how it worked. But it had helped Hare. And he had nothing else to try. So, with the help of a female colleague, Méave Kenny, the Isolation Block's resident medical officer, Colebrook started administering Prontosil to new mothers.

The first patient to receive it from Colebrook and Kenny was a "Mrs. R.," admitted January 14, 1936, six days after labor, with an abdomen "very suggestive of early generalized peritonitis," according to Colebrook's notes. The streptococci had already invaded her abdominal cavity. Her pulse was racing. Her temperature was 104.

Her case was advancing rapidly; it appeared she was going to die. She was given 70 cc of liquid Streptozon—now available as Prontosil solubile, or Prontosil S—both intravenously and intramuscularly. Then they gave her a massive dose of Prontosil, thirty tablets. Colebrook wanted to save her life and administered as much of the drug as he thought she could tolerate. Within hours her temperature began to drop. The infection subsided. A week later she seemed fine. Colebrook, ever wary of overenthusiasm, noted in his journal, "Seems likely that drug may have helped."

This and Hare's case were important observations, but Colebrook was not about to let a pair of anecdotal cures color his thinking. He tried the drug on six more new mothers with childbed fever. One was a twenty-eight-year-old woman with a fast-moving infection, already well established and threatening to spread to the blood. Her fever was peaking. Her chance of survival, Colebrook felt, was less than fifty-fifty. Colebrook and Kenny gave her, too, a very large dose. The next day her temperature dropped. They kept giving her brick-red tablets. Twelve days later all signs of the infection were gone. She was discharged and went home with her baby. The list of successes grew.

At the end of February, Colebrook finally delivered his first internal report on human tests to the Medical Research Council, soberly summarizing what he had seen with the seven women treated so far, plus Hare's hand infection. He framed his findings as conservatively as possible. The hand patient had recovered, he wrote, but "as he did not tolerate the drug well, and therefore received only a small amount, it is doubtful whether there was any beneficial effect." Three of the women (including the cases above) had made unexpectedly rapid recoveries with the use of Prontosil. The fourth, "a fairly severe but not inevitably fatal" case, survived longer than expected, but, Colebrook noted, "there was still no clear evidence that the drug had any effect." The fifth and sixth were doing well, "but it is impossible to say that they might not have done as well without the drug." The fate of the seventh, an extremely grave case, a very advanced infection, was "still in the balance."

Beneath the cautious language there was great excitement. Colebrook might have had doubts about the role the drug played in Hare's recovery, but Hare certainly did not. He believed that Prontosil had cured him. Colebrook's own report indicated that he had seen beneficial effects, perhaps cures, *in every case* in which the drug was used. Colebrook let a bit of his enthusiasm slip out for just a moment in a cover letter accompanying the report. "It seems as if we may be at the beginning of an important new chapter in chemotherapy," he wrote. "In view of the scrappy data given by German and French papers it might be worthwhile when we have done a few more human cases and animal experiments to put out a short paper."

He then started giving Prontosil to every childbed fever patient he could find.

"I was no more than a spectator, but I soon sensed that a change had come," Hare wrote later. "Patients whom we would have given up before now recovered easily, and without the long-drawn-out, desperate illness that would previously have been their lot."

Colebrook and Kenny's "short paper" grew into two long reports in the leading British medical journal *The Lancet,* both published in 1936. The first, in June, began with a nod to Domagk's "startling chemotherapeutic success"; the second, in December, completed and expanded the number of childbed fever cases they had treated with the new drug. Together they form a scientific report of something like a modern miracle.

Before Prontosil, between 1931 and 1935, Colebrook had treated about five hundred cases of childbed fever in the Isolation Block at Queen Charlotte's. One in four of those patients, despite receiving the most advanced medical care in one of the best facilities in the world, had died. Then came the new drug. By the end of 1936, Colebrook and Kenny had treated a total of sixty-four patients with Prontosil. Sixty-one survived. Prontosil cut the mortality rate in childbed fever from somewhere in the 20 to 30 percent range down to 4.7 percent. No hospital had ever seen numbers like that since the first epidemic hit the Hôtel Dieu in the 1600s. Side effects were few (urinary tract irritation notable among them) and relatively mild. Even Colebrook could not restrain a hint of enthusiasm. "The spectacular remission of fever and symptoms," he wrote in the conclusion of the first paper, "does suggest that the drug exerted a beneficial effect."

It was not a perfect study. Once they saw what was happening, Colebrook and Kenny decided that it would be inhumane to withhold the drug from any childbed fever patient. They gave it to every young woman who was infected, without exception. That meant that there were no untreated controls with which to compare results, so they used historical comparisons instead. By today's standards of randomized, double-blind clinical tests, it was a flawed study. However, randomized, double-blind clinical tests were virtually unheard of in the 1930s, when almost no one was doing controlled human testing of new drugs. By the standards of the day, the Colebrook and Kenny papers were enormously compelling and influential. It was the first report of a test of Prontosil on a large number of patients. It came from a trusted source, a respected and conservative physician/researcher, a disciple of Sir Almroth Wright, and a man who had no financial relationship with Bayer; there was no question of touting the drug for commercial gain. And, just as important, the Colebrook and Kenny papers were in English, introducing the new drug to those many physicians in the United Kingdom and abroad who did not keep up with the German medical literature.

The editors of *The Lancet* were so impressed that they accompanied Colebrook and Kenny's work with an editorial comment that seemed equal parts appreciation and perplexity. "The history of attempted chemotherapy in bacterial infections is so discouraging that any indisputable success in this direction is almost totally unexpected," they wrote. "Here there is a substance identified as useful by the method of trial and error, the mode of action of which is totally unknown . . . in this sphere of medicine we are only groping in the darkness and finding something here and there, more by good fortune than by intelligent design." The editors commended Colebrook and Kenny for their work and ended, "It is very much to be hoped that the therapeutic trial that they have initiated here will be extended and made to embrace other types of acute streptococcal infection."

They did not have long to wait. Colebrook and Kenny's studies had an immediate impact, quickly built upon by other clinicians. Within two years, as Bayer expanded the availability of Prontosil commercially first throughout Europe and then the rest of the world, sulfa-containing drugs became a standard treatment for childbed fever. By 1938, it was estimated that sulfa was saving the lives of ten thousand new mothers each year in Britain and the United States alone. Colebrook was lauded in the English press and applauded by physicians around the world. Typically, his response was, "I do wish people would not exaggerate my part. I was always part of a team."

The good news extended beyond childbed fever. One of Colebrook's assistants on a side mission, trying to determine if infectious strep could survive in dust, was poking around collecting samples in the cupboards and corners of Queen Charlotte's when he apparently breathed in enough bacteria to incur a bad case of strep throat. This was not exactly the way he had hoped to prove his point, but it had its value. When the resultant tonsillitis threatened to release strep into his blood, Colebrook gave the assistant a large dose of Prontosil. "It was absolutely terrible," the man recalled later. "Almost immediately I had the most awful feeling in my gut. I could almost feel it going around my bloodstream, very hot. . . . I thought I was going to die." Violently sick, he was put to bed with the childbed fever victims in the Isolation Block, a lone man among thirty or so pregnant women.

There he, too, quickly recovered.

IN LATE 1936 Sir Henry Hallet Dale was informed that he had won that year's Nobel Prize for physiology and medicine. The prize was to be shared, however, with a longtime colleague in the work on nerve impulses, Otto Loewi, a German-born professor who did research at the University of Graz in Austria. The joint award of the prize to scientists in Britain and Austria was significant. Republicans and Fascists were battling in Spain; Adolf Hitler was rearming and building the German air force in clear violation of the Versailles Treaty; concern was growing throughout Europe that events were leading to another world war. But science, it seemed, still placed knowledge over nationality. Prontosil was a good example: a powerful new medicine developed in Germany, tested in England, promising to ease human suffering throughout the world.

Science still seemed to offer reason for optimism. By the time Sir Henry accepted his Nobel medal in Stockholm, however, Hitler had marched troops into the demilitarized Rhineland, Japan and Italy had allied with Germany—and the French, just four years after the discovery of Prontosil, were about to make the new German miracle drug worthless.

CHAPTER THIRTEEN

Ernest Fourneau, a leading French chemist, did just as Sir Henry Hallet Dale had done immediately after reading Domagk's work in the February 1935 *Deutsche medizinische Wochenschrift*—he wrote Heinrich Hörlein to request samples of Prontosil. Fourneau, however, did not expect to receive them anytime soon. Fourneau knew Hörlein, Hörlein knew Fourneau, and they both knew what the French scientist was likely to do when he got hold of the new drug: study it, solve its structure, devise some new way to make it, then give it to French drugmakers who would compete with Bayer. In other words, from Hörlein's perspective, Fourneau would steal it.

A thin, elegantly dressed, sophisticated, and respected intellectual, Fourneau had been competing with the Germans for decades. He headed the pharmaceutical research unit at the Pasteur Institute in Paris, dedicated, like Hörlein's program, to finding new chemical medicines. But where Hörlein controlled an entire factory with armies of technical workers and enormous sums of research money, Fourneau ran a small lab in an old building in an unfashionable district of Paris. The odds seemed uneven. Fourneau, however, made up in cleverness what he lacked in resources.

Fluent in German, French, and English, habitué of literary salons, lover of fine art, his white goatee impeccably trimmed, his urbane appearance marred only by a slight limp and a bad eye—marks of honor, both the result of laboratory accidents—Fourneau was one of the world's foremost jugglers of molecules. He had learned how to out-think the Germans from the Germans themselves. He had studied in Germany for three years, admired and envied their successes, and wanted the same sorts of successes for France. So he studied, watched, and, wherever possible, profited from German advances in medicine. When he saw the Prontosil articles, he knew immediately that Hörlein's group was onto something big. So he sent his request for samples to Hörlein and waited.

Not that he expected a quick reply. His relations with Hörlein were, to say the least, strained, and had been ever since Fourneau had found a way to duplicate Bayer's Germanin molecule, the sleeping sickness cure found by Roehl, and then gave the formula to a French drug company to produce under another name. It was all legal according to the patent laws of the two nations, but Hörlein had never forgotten the insult. It was unlikely that he would fill one of Fourneau's requests for an experimental drug very quickly.

Fourneau could wait. He knew how the German mind worked. He had admired the Germans ever since he was a young boy, first through their music—Fourneau was a talented pianist who had studied with a German teacher and grew to love German composers—and then through their culture. His family had not been extraordinarily rich, but Fourneau was comfortable around the wealthy; he was raised in the exclusive seaside resort town of Biarritz, where his family owned a hotel. He spent his youth hobnobbing with affluent hotel guests, including a German count, a diplomat who made a great impression on eighteen-year-old Fourneau, teaching the boy about German culture, German history, and the German soul. Fourneau read Goethe and Schiller, played Beethoven and Bach. When he switched his interests to science, he traveled to Germany and spent years studying chemistry in the laboratories of two future Nobelists, Emil Fischer and Richard Willstätter. Here again he was impressed by the German mind.

His heart, however, was resolutely French. He returned to France and took jobs in the drugmaking industry, starting in 1901 as research director for the nation's largest drugmaker, Poulenc Bros. At that time the French drugmaking industry was "virtually nonexistent," he remembered. "Everything came from Germany, or was manufactured in German plants installed in France." Every time a French firm tried to run competition by building a factory to make a new substance, the Germans—who had already paid for their factories and so had lower overhead costs—would lower their prices and drive them out of the market. Fourneau decided that if the French were going to compete, the nation's scientists would either have to discover their own new drugs and get them into production before the Germans could or find ways to make French versions of German compounds before the Germans had earned back their research and production costs—in other words, get French versions of new German drugs into the market before the Germans could lower their prices. French patent laws, like those in Germany, did not protect the final product. "I was always against the French law and I thought it was shocking that one could not patent one's invention," Fourneau said, "but the law was what it was, and there was no reason not to use it."

Use it he did. He read every German journal in medicine and chemistry, studied business publications, tracked the patent listings, and spent eighteen hours a day compiling a master file of French pharmaceutical needs, German exports, and French imports. Every year, instead of vacationing, he attended German scientific congresses and the trade exhibitions that followed, where he could see the latest products and equipment.

France was a backwater by comparison. French medical chemistry was less an industry than a glorified hobby. The model was Louis Pasteur, a hero of the people, a selfless chemist working for the most part on his own, saving individual lives, taking a personal interest.

Fourneau was determined to bring the best parts of the German approach to France. After starting at Poulenc, he quickly discovered a synthetic alternative to cocaine, widely used (and increasingly abused) in medical procedures as an anesthetic. Fourneau's new molecule, Stovaine (a play off his own name; in French, *fourneau* means "furnace"), brought the young Frenchman to the attention of the great Duisberg himself, who offered him a job running Bayer's research operation in France. Despite the munificent salary Duisberg dangled in front of him, Fourneau turned it down.

After the excitement of Ehrlich's introduction of Salvarsan, Fourneau accepted an offer to start the first laboratory for chemotherapy at the Pasteur Institute, married into an aristocratic Parisian family, and began implementing his ideas for competing with the Germans. Just after World War I, he even did a bit of spying, acceding to a request from the French war minister that he travel to Germany, review German industry, and report privately on the status of chemical research. He visited Bayer, was struck by the size of its humming factories and impressed by the investment Duisberg was making in research. Fourneau realized that Bayer was funneling a fortune into new-product development to make up for its war-related losses.

There was no way for the Pasteur Institute to compete—at least not directly. The two operations differed in more than sheer size. Bayer was private industry, the researchers all male, the chemists generally German, the work often secret, and the goal clearly profit. The Pasteur Institute was more a diverse enterprise based on the selfless model of the great founder, whose remains were then (and still are today) enshrined in a marble crypt in the basement of one of the institute's buildings. Pasteur worked for the good of mankind, not for profit. His institute attracted bright young idealists from around the world, who worked in a community of scholars rather than a factory for discovery. Many of them were women. It was a collaborative, cooperative, relatively open atmosphere; there were romances at the Pasteur Institute, and gossip, personal enmity, cranky individualism. Marriages were proposed. At least one Pastorien (as they called themselves) challenged another to a duel.

Fourneau's chemotherapeutic laboratory comprised a typically talented, polyglot group. It included a rarity in the world of chemistry in those days, a handsome married couple, Jacques and Thérèse Tréfouël (Jacques was Fourneau's number-two man, the manager who ran the laboratory day to day; Thérèse a skilled bench scientist). There was Federico Nitti, a son of a former prime minister of Italy who went into exile when the Fascists took over. There was Nitti's sister Filomena, who was in love with another of Fourneau's group, Daniel Bovet, a brilliant Swiss chemist who would later win the Nobel Prize. It was, one member recalled, a wonderful place to work, a "warm-hearted laboratory." Everyone who worked for Fourneau admired him and respected him, although he was not a typical Pastorien. Longtime laboratory directors and higher managers at the institute were often affectionately called "Père" (father) by the younger investigators. Fourneau never was. He was always "the Master" when his lab workers talked among themselves and "Monsieur Fourneau" to his face.

"The Master would often come to the counters and scrutinize all things with a look devoid of indulgence," one of them wrote. "He would sometimes take a flask from the chemist's hands, take out of his pocket a glass agitator he always carried, then would accelerate the reaction by heating, cooling, agitating, his right arm in the air, scrutinizing the mixture. Often the reaction would occur and, without a word, Fourneau would put the container down and leave." Even though he never raised his voice, he was able to exert total control through his commanding presence. Sometimes the Pasteur Institute would send him problem employees, whom he had no trouble managing.

The relatively limited funding available at the Pasteur Institute did not stand in the way of discovery. Fourneau's lab was the first to make an oral arsenical, for example, for use against syphilis. But an occasional success did not mean that his team was equipped to compete with the Germans in finding new medicines. So they worked on Fourneau's other strategy, that of finding French versions of new German drugs and getting them quickly to the marketplace. Their first success came in 1925, when they broke the formula for Germanin. Bayer moved swiftly to repair the damage by initiating joint sales efforts with the French, but the company lost at least part of a lucrative market—and Hörlein lost whatever trust he had for Fourneau.

So when Hörlein received a letter from the Pasteur Institute requesting a sample of Prontosil "for experimental purposes," he replied positively but asked in return for a meeting in which he and Fourneau could discuss sales and marketing of the new medicine in France. The meeting was held, the participants were polite, but nothing was resolved. The French went to work deciphering what they could of the German patent application for Prontosil. Very soon they found a way to replicate the molecule, not precisely, but closely enough to work. Constantin Levaditi, a longtime Pastorien, confirmed that the French chemical worked in animals almost as well as the Germans reported for theirs. It all happened very fast. Levaditi published his results within three months of Domagk's paper, and the Pasteur Institute began distributing the drug to French physicians, who reported "spectacular results" using it against erysipelas and other streptococcal diseases. The recipe went to a French chemical firm that quickly began manufacturing and selling it under the trade name Rubiazol, for its red color. When he heard the news, Heinrich Hörlein started fuming. He had spent eight years and millions of marks discovering, developing, and bringing Prontosil to market. Within three months of its announcement, Fourneau had robbed him of at least some of his profits.

Fourneau's group continued to study the drug. In July, Jacques Tréfouël gave Bovet and Nitti the task of testing Prontosil/Rubiazol as well as a series of the Pasteur Institute's own chemical variations for effectiveness against strep in mice. Between July and November 1935, working around the usual summer holiday (during which the institute virtually shut down), they perfected their animal-test system. Bovet and Nitti found, as Domagk had found before them, that much in mouse experiments depended on the particular strain of *Streptococcus* being used; they got variable results until they started using a particularly vicious strep culture they obtained from a victim of childbed fever at the Hôtel Dieu. Against this germ they tested eighty chemicals, including many azo compounds. Their findings fully confirmed the German results.

Then came something entirely unexpected. "It happened on November 6, 1935," Bovet wrote. "We received a large number of mice (forty) and placed them in glass jars in groups of four. They all received an intra-peritoneal injection of highly virulent streptococcal culture. One group was kept as a control group and the other received a sufficient dosage of the coloring agent described by Bayer [Prontosil]. After that we treated seven groups orally with the products we had synthesized in the lab. But I only had seven new products and we had an extra group of four mice. Why, I asked, not just try the product common to all these products, para-amino-phenyl-sulfonamide?" The long technical string at the end of Bovet's sentence is a chemical name for pure sulfanilamide, the side chain the Germans used to awaken their azo dyes.

Bovet's casual decision would change the history of medicine. Everything the Germans had been testing at Bayer involved linking sulfa to something else, almost always an azo dye. Everything the French had been trying was done the same way. Testing pure sulfa alone was something the Germans had nosed around—they had tested molecules that were close to pure sulfa, yet they always appeared to just miss the exact molecule itself—but never quite accomplished. Then came four extra mice and a bit of handy chemical. Bovet and Nitti separated the groups of mice; marked them with stains of picric acid, a yellow spot on the tail for one group, patterns of yellow spots on the sides or back for others; infected them with strep; injected them with chemicals; and observed them on their bedding of dried oats. Each dead mouse received a cross in the lab journal, each one who lived got a *V*. The next day the control group, the unprotected mice, were dead or dying, as expected, as were almost all of those treated with the experimental chemicals. Only one of the new French azo dyes seemed to be working. Those mice treated with Prontosil were alive, also the expected result.

The surprise came in the last jar, which contained the four extra mice, the group treated with pure sulfanilamide. They were not only alive, Bover recalled, but were "doing great." It was one of those results that no one quite believed until retests proved the point—retests that Bovet and his coworkers performed immediately after informing Fourneau. It was difficult to accept, but it appeared to be true: Simple sulfanilamide, a colorless, common, unpatentable, off-the-shelf chemical used by the pound in the dye industry, was as potent a medicine as the German wonder drug Prontosil. The implications were immediately clear to the bright young men and women in Fourneau's laboratory. "As we wrote the last 'V' on the report, we had already realized that the future belonged to 'colorless products,'" Bovet wrote. "From that moment on, the German chemists' patents had no more value whatsoever."

The finding meant much more than money. It offered explanations, opening the way to understanding why attaching sulfa to many types of azo dyes resulted in an active medicine, while dyes without sulfa were far less active, if they had any medicinal value at all. It pointed toward an answer to the mystery of why Prontosil worked in live animals but not in the test tube, with the French team immediately hypothesizing that sulfa had to be released from the rest of the Prontosil molecule in order to become active. That could happen in the body of an animal, where enzymes in the body could split the Prontosil molecule into two pieces, releasing pure sulfa as the medicine. The dye did nothing but stain the skin. The sulfa cured. In the test tube, there were no enzymes to split the Prontosil, and the sulfa would not be released. This discovery, which the French explored and confirmed, opened up an entirely new field of medicine, in which substances could be "bioactivated" in the body.

Most important, however, was the discovery that the world's most effective antibacterial medicine was also among the simplest ever found. Everyone had been searching through all these complicated dyes, tinkering around the edges, while the real power was in a colorless add-on. As Bovet later put it, the Germans' complicated red car had a simple white engine.

Fourneau was elated, but he refused to take credit for the finding. A few days after confirming their first experiment, Bovet and Nitti brought him a copy of a report they had drafted for publication. They expected that he would accept his name where they had placed it, at the top of the list of authors. He surprised them by crossing his name off entirely. "The reasons—or maybe the feelings—which made Fourneau do that were hard to understand for those close to him," Bovet wrote. "Was it a generous gesture towards younger colleagues, in order to help their careers? Or was he being deferential to and respectful of H. Hörlein and the scientific staff of Bayer, and worried he might hurt their scientific prestige by this publication?"

The work appeared under the names of the Tréfouëls, Bovet, and Nitti, in the *Comptes Rendus des Séances de la Société de Biologie* (Reports of the Society of Biology) at the end of 1935. They kept it brief, not because they thought their paper was unimportant but because they knew that the secretary of the Société de Biologie, in order to pack in as many communications as quickly as possible, measured each submission with a centimeter ruler and discarded whatever was too long.

The French team then set to work digging further into their discovery. Now that they knew where the active portion of the molecule was, they could move away from dyes entirely and attach sulfa to many different side chains at different points, then measure the effects. They were looking for ways to make sulfa stronger or safer and also for clues into how it worked its magic. They made many different sulfa-containing compounds that varied terrifically in terms of chemical personality—solubility, surface tension, melting point—but many of which stopped strep. If the sulfa portion was lost or altered, however, the effect went away. They made a number of variations on the sulfa molecule as well, moving pieces around, seeing what happened. All of their results were quickly published. Secrecy was not an issue.

It was wonderful that this powerful, inexpensive medicine was now available, but for a year after the Pasteur Institute announcement, no one marketed it seriously in its pure form as a medicine. Because it was not patentable, it was difficult for major chemical or drug firms to see a way to make much of a profit from it. It was not until months after the Pasteur group's first publication on sulfa that the president of Rhône-Poulenc, an industrial supporter of Fourneau's laboratory, visited the Pasteur Institute to hear about it. After talking with the researchers he decided to launch Septazine, a variation on pure sulfa that he felt was different enough to allow patenting—and hence profits. Septazine reached the marketplace in May 1936. In June a British group confirmed the French findings. There was still no word from the Germans. "We at the Pasteur Institute were extremely anxious to find out the reaction of the Bayer researchers to our discovery," Bovet said. But no reaction was forthcoming—at least not publicly.

CHAPTER FOURTEEN

As soon as it appeared, the first paper by the Pasteur Institute team was translated into German and circulated within Bayer. According to one of the German chemists, it "struck like a bombshell. . . . The excitement caused by this paper can only be appreciated by those who witnessed it." The revelation that the sulfa side chain was the active ingredient in Prontosil rather than the dye set off a string of responses that started with denial: Certainly there had to be something wrong with the French results. After Klarer and Mietzsch provided Domagk with pure sulfanilamide and his tests showed that it worked—not only worked but, on a gram-for-gram basis, was twice as effective as Prontosil—the Bayer team moved to confusion: Mietzsch and Klarer in particular were stunned. How could the medicine work if it was not a dye? This undid everything Ehrlich had taught. Then came recrimination: How, after three years of research, had they missed a critical discovery that the French had made within a few weeks? Then finger-pointing: At least one observer within Bayer noted a "confrontation between Mietzsch and Klarer on the one hand and Domagk on the other," in which, presumably, each side tried to assign responsibility for the embarrassing French finding.

Finally the Germans found their way to acceptance. Something had gone wrong, their group had executed a major scientific pratfall, but it was done, and they had to deal with the results. Who was to blame? No one and everyone. The chemists had provided Domagk with hundreds of sulfa-containing chemicals, almost all of which worked. Within three months of discovering Prontosil, they had sent him Kl-820 and Kl-821, molecules in which sulfa was not attached to an azo dye. Domagk, for his part, believed that he had run his tests flawlessly. Almost every time they tested an azo dye with a sulfa side chain, it killed strep; almost every time it did not, the effect was absent or greatly reduced.

Perhaps there were enough confounding results—tests in which an occasional azo dye without sulfa proved somewhat effective, usually on bacteria other than strep; tests in which, for no apparent reason, a sulfa-containing azo dye that should have worked did not—to throw them off, keep them focused on azo dyes instead of sulfa.

Because many documents at Bayer (including all records of upper-level administrative deliberations) are not publicly available, it is impossible to know exactly what happened. In hindsight it appears that the German laboratory data pointed clearly toward the French results and that the Germans, at least to some degree, knew it. The Bayer chemists had followed their intuition and, just like the French, within months of discovering Prontosil had developed molecules like Kl-820 and Kl-821, compounds that contained sulfa but no azo dye. They gave the molecules to Domagk, Domagk tested them, and a confirmed, positive set of results from at least one of these non-dye molecules was ignored, dismissed, or buried within the company. The team seemed well on their way to finding what the French found. Then they stopped.

Or did they? As early as November 1932, within weeks of first making the molecule that became Prontosil, Klarer in his internal reports was already referring to the sulfa side chain as "the active group" and adding that Mietzsch was helping him to produce sulfa in large quantities. Weeks before submitting the patent for Prontosil, Klarer noted that he had prepared at least two compounds in which the sulfanilamide group was attached to non-azo molecules, adding that he planned further work on them. Domagk's laboratory notebook includes no mention of any tests on those substances.

In the fall of 1934, more than a year after testing the non-azo sulfa compound Kl-821 and more than a year before the French results, Klarer and Mietzsch sent Domagk another non-azo chemical, Kl-1123, a molecule that was almost, but not quite, the pure sulfa the French later tested. Kl-1123 differed from sulfa by only three atoms, a grouping of a single nitrogen and two hydrogens connected to sulfa's benzene ring. It appears to be an attempt to test the sulfa side chain alone. Domagk ran Kl-1123 through his standard battery of tests in early September 1934—and found it completely ineffective. To work well, as the French later found, the sulfanilamide molecule had to be built just right. A change as small as three atoms in the wrong place could render it powerless. The Germans had turned their attention back to dyes.

Why the extra three atoms? Once he saw the French results and realized that sulfa alone worked, Domagk might well have asked why his chemists had given him a molecule just off the mark. The chemists could have accused Domagk of messy animal testing in which negative results for sulfa-containing dyes had thrown them off the path. There is no way of telling.

One thing is certain: The German team at Bayer was entranced by azo dyes. Klarer and Domagk both were struck by the results they saw in which azo dyes that contained no sulfa showed effects against diseases as diverse as tuberculosis and rat leprosy. Azo dyes had led them to Prontosil, and azo dyes continued to capture their imagination afterward—despite all the signs pointing to sulfa.

Whatever the specifics of the arguments between the chemists Klarer and Mietzsch and the pathologist Domagk in the winter of 1935, they were quickly quieted by Hörlein, who stayed calm, told the chemists and the pathologist to go back to their neutral corners, and ordered further tests. Hörlein refused to assign blame. He needed his team to function.

He had more important things on his mind, notably explaining the fiasco to his superiors at Bayer. The company was pushing Prontosil into an expanded phase of global distribution. If a simple, unpatentable bulk chemical was as powerful as the antibacterial dye that the company had spent years developing, their work could be rendered worthless. After discussions at the highest levels, the decision was made to proceed with the Prontosil introduction while further tests based on the French results continued in Domagk's laboratory. There was still a good chance to pull out of this with money. Many factors pointed toward Bayer's ability to profit from its investment despite the French findings. The Germans were first into the field with a proven, tested, drug. The Bayer name meant quality, and Prontosil would benefit from that association. Physicians were already using the drug and publishing results that referred to Prontosil by name, a valuable form of free publicity. Demand was growing, and Bayer had both the product and the worldwide sales mechanism to meet it. So Bayer continued with its rollout of Prontosil—while at the same time starting to make, package, and sell pure sulfa, the French discovery, under the trade names Prontosil album ("white Prontosil"), Prontylin, and Prontalbin, all names designed to point attention back to the original Bayer discovery. For a while the strategy worked beautifully. Sales were healthy. For more than a year, the Prontosil family of drugs was second only to Aspirin as Bayer's top-selling medicines.

Bayer said nothing publicly about the French results. The Pasteur team waited in vain to hear how their work was being received. A year passed before Domagk told a meeting of the Society of German Naturalists and Physicians that the "raw material" of Prontosil, sulfanilamide, had a similar therapeutic effect when administered orally. He mentioned that it was available from Bayer as Prontosil album. He made no reference to the work at the Pasteur Institute.

Publicly, Hörlein was calm and quiet. Privately, he was incensed. When Fourneau had first requested samples of Prontosil, Hörlein himself had gone to Paris to sit with him and a representative of Rhône-Poulenc, the French pharmaceutical firm, to discuss how the new German drug would be marketed. Hörlein had tried to get an agreement on malaria drugs as well. He left feeling that they had made at least some progress. Then came the announcement that the French had discovered how to make something close to Prontosil on their own. That was quickly followed by the news that Fourneau's lab had discovered the power of pure sulfa, threatening everything Bayer had put into Prontosil. That, Hörlein complained to Fourneau, "made me upset, I admit, because I had at least expected that you would contact me before preparing a publication in a field that we had opened." The French paper had caught him "completely unprepared," he wrote, adding that his team, too, had been working on colorless sulfa drugs—a surprising admission given their failure to find one.

Hörlein's pride had been hurt, and he intended to get his team back to the front of the field. Fourneau's lab continued to study and publish work on pure sulfa through early 1936, as Hörlein, Domagk, Mietzsch, and Klarer tracked, translated, and shared the stream of publications in French medical and chemical journals. The French work now extended to clinical reports from hospitals where French physicians were finding that pure sulfa could cure a variety of strep diseases, notably erysipelas and blood infections. Late 1935 and early 1936 were great times in the Fourneau lab. The French team was dancing around IG Farben, outpublishing and outperforming the Germans at every step.

The only difference was that Bayer, now introducing Prontosil worldwide, was making all the money. Hörlein was also able to throw money at the problem, expanding his research team with a new chemist and more assistants for Domagk, and focusing everyone on the next steps: Use the French results, forget about the azo core, and find new sulfa compounds more soluble, even less toxic, and more widely effective against more diseases than plain sulfa—and, of course, make certain that these compounds would be patentable as well. The Bayer chemists started pumping out new molecules at such a rate that a bottleneck formed. There were so many chemicals to test that Domagk was running out of appropriate laboratory animals. The lack of mice became an important factor slowing Bayer's development of next-generation sulfa drugs.

ON THE LAWN of Hermann Göring's palace next to the Ministry of Air on the Wilhelmstrasse in Berlin was gathered a glittering array of the aristocracy of Germany, old and new, enjoying conversation, champagne, and the fine weather of a summer afternoon. It was 1936. Ernest Fourneau found himself, almost against his better judgment, enjoying himself. He had come to attend the 1936 Olympics—"except for a few extraordinary Negro runners, the Germans won everything," he wrote—and had been invited to a number of other events surrounding the contests. He attended concerts and chatted at formal dinners. He saw a huge Nazi rally in Nuremberg and heard Hitler speak. He had been impressed.

Now he stood on Göring's lawn, elegantly dressed in evening clothes, enjoying himself and thinking after a few glasses of champagne that his host was, after all, an extremely agreeable man, presented himself well, with the head of a Roman emperor and a voice strangely soft and pleasing. Pleasing, too, were the partygoers, especially the old aristocracy, the princes of the blood—the emperor's own son was there—whom Fourneau saw speaking privately, animatedly with the leaders of the new Reich. This was quite unlike what he had expected from the Nazis. It was so different from the situation in France, where the old royalty were pushed aside. He had always admired the Germans. And now he admired them even more. "All were joyful and let themselves enjoy the charm of this unique soiree," Fourneau later wrote. "Those whom hatred does not blind, whose soul is not undermined by skepticism, those who are still able to dream and whose heart has remained young, they understood, when they watched the Olympic Games in Berlin and went to the receptions that followed, they understood that a new world was rising before their eyes, a world that tried to reconcile the most intelligent and most efficient love for the people and for youth with a taste for art and luxury, with the utmost refinement, and the respect of traditions."

Fourneau enjoyed cooperating with Germans almost as much as competing with them. He had visited Germany often in the 1920s and 1930s, nightclubbing in Berlin, enjoying Wagner's great music at Bayreuth, motoring through Bavaria (where he noted in 1932 "many groups of unemployed people of both sexes, holding guitars and accordions, living a life of complete freedom, sleeping in the woods, the farms, the castles turned into youth hostels"). He also saw his first SS members in 1932, "young, tall, very well dressed," always in pairs, quietly monitoring Munich cafés and pubs.

The Nazis cultivated men like Fourneau, invited them to events, and dazzled them when they arrived. Fourneau's continued enthusiasm for German science and culture worked in the Reich's favor. The French scientist had a history of open-minded friendliness toward the Germans. After World War I, Fourneau had helped develop a formal system for allowing French and German science students who started their studies in one nation to finish them seamlessly in the other. He attended German scientific congresses and meetings, joined a Franco-German cooperation committee, and made a point of bringing German scientists to France. Rapport with Fourneau was a way of showing the world a Germany reaching out to a former enemy nation.

IG FARBEN, too, was learning to get along with the Hitler regime. In the early days, the days of the first elections and Hitler's accession in the early 1930s, the giant firm's attitude toward the Nazis had been cool at best. But Farben's income depended on worldwide sales, and worldwide sales depended on stable relations with other nations; the company found it most profitable to steer clear of partisan politics, to support both sides in elections, and to accommodate to whatever government came to power. Industrial leaders like Bosch might have felt that the Nazis were odious in many ways—he met personally with Hitler to protest Nazi policies toward Jewish scientists and received a heated rebuke in reply—but they learned to get along. Like most large businesses, Farben was inherently conservative. The Nazi promise of a stronger economy and a stronger central government were welcome. The group's social policies were another matter, very troubling, with a tendency toward economic and social experimentation, an aggressive foreign policy, and that antipathy toward Jews—a group that included a number of IG Farben's best scientists. With few exceptions Farben's top officials generally did not support the Nazis in elections before 1932. But neither did they come out strongly for the party's opponents. The company took a wait-and-see stance. After Hitler became chancellor, IG Farben found it most advisable to get along with the new government. There were positive aspects to the Nazi administration. For one thing, the Nazis promised to funnel state support into Farben's enormously costly gamble to make artificial gasoline, a huge project badly in need of an infusion of cash. Hitler seemed in general to support scientific research, and despite the exodus of Jewish researchers from Germany after 1933, many non-Jewish researchers and doctors—some 60 percent of biologists, 80 percent of anthropologists, and large numbers of physicians—joined the Nazi Party during the decade. After Hitler came to power, Farben's leaders decided simply to ride out the Nazi era, use the government where possible, and concentrate on business. As it turned out, the Nazis would use Farben.

By the mid-1930s there was change everywhere in Germany, as the Nazis consolidated their control and continued rebuilding the military. There was change, too, at Bayer. Carl Duisberg, the great industrialist, the company's father, died one month after Domagk's first publication, in March 1935. When he was buried, the giant Bayer factory on the Rhine was closed for the day; the entire town of Leverkusen turned out to honor him. It was the end of an era.

Through all the changes, Heinrich Hörlein managed to keep his drug research operation moving more or less smoothly. Hörlein agreed with many things the Nazis stood for. His feelings about Jews, for instance, were well known. He hired as few as he could, and he had helped both to "cleanse" Jews from the board of the German Chemical Society in 1933 and to get Jewish editors removed from the society's professional journal. In other ways he worried about the Nazis. He was not happy with the party's condemnation of animal experiments, which Nazi leaders had ignorantly spoken of as a form of "Jewish science" linked to ritual slaughter. To Hörlein, animal experiments were both good science and an absolutely essential part of his work. He sought out a party official who might help him dissuade Hitler from following through on promises to end the practice, finally persuading one of Ernst Röhm's top men, Julius Uhl, to pay a visit to Elberfeld. The storm trooper official toured the facilities and heard about the importance of animal tests in ending disease. Everything seemed to go well. Then, a few weeks later, Uhl publicly denounced IG Farben for developing and selling tropical medicines that benefited the usurpers of Germany's former colonies, saying, according to Hörlein, "that he [Uhl] wished to have nothing more to do with us, that we were not only an international Jewish undertaking, but simply traitors . . . [and that] in the future, we should be dealt with in quite a different manner." So this was how the Nazis worked.

In June 1934, Hörlein, always eager to keep things running smoothly, joined the Nazi Party.

GERHARD DOMAGK kept his political opinions to himself. He had plenty of other things to think about. In early December 1935, just after the French published the discovery that pure sulfa was the active ingredient in Prontosil, Domagk's six-year-old daughter, Hildegarde, suffered a bad accident. She was making a Christmas decoration upstairs in their house when she decided that she needed help threading a needle. She was on her way downstairs to find her mother, carrying the needle and thread, when she fell. The needle was driven into her hand blunt end first, breaking off against a carpal bone. She was taken to the local clinic and the needle was surgically removed, but a few days later, her hand started swelling. After the stitches were removed, her temperature rose and kept rising. An abscess formed at the surgical site. She had a wound infection. The staff at the clinic tried opening and draining the abscess. When it became reinfected, they opened it again. Then again. The infection started moving up her arm. "Her general state and the abscess worsened to such a point that we became seriously concerned," Domagk wrote later. "More surgery was impossible." She was falling in and out of consciousness. The surgeons were talking about amputating her arm. Once the blood tests showed that the invading germ was strep, Domagk went to his laboratory and pocketed a supply of Prontosil tablets, returned to her hospital room, put the red tablets in her mouth himself, and made certain that she swallowed. Then he waited. A day later her temperature continued to rise. He gave her more tablets. No improvement. On day three he gave her more, a large dose, but there was still no improvement. Her situation was growing desperate, so he pulled out all the stops, on day four giving her more Prontosil tablets, then two large injections of Prontosil solubile. Finally her temperature started to drop. He gave her more tablets. After a week of treatment, her temperature finally returned to normal. The infection had been stopped. By Christmas she was able to celebrate the holidays with her family.

A full year after the French published their results, the German patent on Prontosil was anything but worthless. Bayer had introduced its family of new medicines to the market in Germany, then throughout Europe, with great success. The company was slower to roll it out in America. Bayer had applied for a U.S. patent for Prontosil back in 1933, but in the summer of 1936 it still had not been granted (and would not be until the summer of 1937). In the absence of a firm patent, the company was keeping a low profile in the United States. A few American physicians, those who read the German journals or had contacts in the United Kingdom, had heard about Prontosil; a few had even tried it. For the most part, however, in mid-1936 the United States was virgin territory for sulfa. That was about to change.

It started in London. There, on July 27, 1936, Leonard Colebrook gave an address to the Second International Congress of Microbiology about his work with childbed fever. It was impossible for even the congenitally reserved Colebrook to hide his enthusiasm. The drug clearly cured these infections, and it was well tolerated. Physicians and bacteriologists from around the world who were still scoffing at the German claims and paid little attention to the French became convinced that there was something to Prontosil.

Among those attending the conference were two U.S. physicians from Johns Hopkins, one of that nation's leading medical institutions. Perrin Long and Eleanor Bliss were young, enthusiastic, and experienced in the ways of strep infections. They had been working together for years in a vain attempt to make an effective anti-strep serum. Bliss had been born into one of the leading families in Baltimore, graduated from Bryn Mawr, then chose medicine over the Social Register. She was an expert in animal experimentation. Long focused on patients. He was the restless type: born in Ohio, ambulance driver in World War I, awarded the Croix de Guerre for bravery in action, back to the United States for his M.D., back to Europe to serve as a voluntary assistant in a German hospital, back to the United States to join the medical faculty at Hopkins, now in London for the Congress.

Colebrook's address was the grand finale of a well-attended session focusing on new therapies for strep infections. The Tréfouëls were there, too, summarizing the sulfa research ongoing at the Pasteur Institute. Long delivered a paper himself on the serum work he and Bliss had been doing at Hopkins. Then came Colebrook's talk on childbed fever. Everyone who attended the session came away appreciating that Prontosil was everything the Germans were claiming and more. Long and Bliss were invigorated by what they heard. After all their work, it was now clear that there was no point in sticking with serum. What they needed to do was get back to the United States and start testing Prontosil as quickly as possible. Long canceled his plans for further travel on the Continent, then cabled the pharmacist at Johns Hopkins and told him to beg, borrow, or steal some Prontosil or plain sulfa, whatever he could get his hands on. He and Bliss caught the next boat home. When they got there, they found no Prontosil. It was still difficult to obtain in the United States. A laboratory at DuPont, however, was able to deliver ten grams of pure sulfa, enough to get them started. Bliss tried it in mice first. She must have been using the right strain of strep, because she quickly achieved "astounding" results: Eight of ten treated animals survived a strep infection that killed every control animal. "We have never seen anything like this before," she wrote. Long, with the benefit of knowing Colebrook's findings on doses and safety, moved immediately to patients, starting with the worst cases. There was a seven-year-old girl with severe erysipelas, skin flaming and a fever of 105. The doctors had tried everything they could—transfusions, serum therapy—and nothing had worked. By now the Hopkins team had finally secured some Prontosil, and Long began giving the girl the drug every four hours. After two doses her rash started retreating. After three her temperature returned to normal. "It was hard to believe what our eyes had seen," Long and Bliss said. There was a woman suffering from an infected pelvis after a botched abortion. Prontosil saved her within seventeen hours. A two-year-old boy with erysipelas. Cured. A child with scarlet fever. Cured. A young woman with acute tonsillitis. Cured. As fast as they could get Prontosil, they started using it on every strep disease they could find, treating scores of patients with childbed fever, middle-ear infections, blood poisoning, quinsy, peritonitis, impetigo. There were scattered reports in Europe of the drug's working on most of these diseases, but nothing like the Hopkins tests; all those patients with all those conditions treated at a single place by a single team of physicians, had ever been done before. News of Long and Bliss's success raced through the medical center. Other Hopkins physicians began trying the new drug, some launching studies to figure out how it worked, others giving it to strep victims and, as as a result, a popular magazine later enthused, "snatching patients out of the grave."

Long and Bliss might not have been the first in the United States to use the new drug, but Johns Hopkins was by far the biggest, most enthusiastic, most influential U.S. center using it in 1936. Hopkins was respected around the world. Hopkins researchers published their work in leading journals. Hopkins did more than any other center in seeking to solve the mystery of how the drug worked. In addition to Long and Bliss's clinical and animal results, for instance, E. K. Marshall at Hopkins discovered a simple way to measure the concentration of the drug in body fluids, leading to more effective ways to relate dosage to effect. By the fall of 1936, Hopkins researchers were in the middle of a detailed comparison of Prontosil with pure sulfa, confirming in a more carefully quantified way from metabolic and toxicological studies what the French were saying: Both worked.

Sulfa was such a hit at Hopkins that other physicians started joking about it, saying that if you went to see a doctor there, you were immediately put on sulfa; if you were not well after a week, they might give you a physical exam. Long got calls from physicians pretending to be famous people asking for miracle cures. One day he was with a colleague when his phone rang. The other physician answered and heard a woman's voice on the other end ask for Dr. Long. He gave Long the phone and heard him say, "You can't fool me this time. I know you're not Eleanor Roosevelt." Then Long hung up. The phone rang again a few seconds later. This time Long answered himself. After a moment he said, meekly, "Yes, Mrs. Roosevelt, this is Dr. Long."

CHAPTER FIFTEEN

Franklin Delano Roosevelt Jr. was having second thoughts about his engagement.

Not about the girl. His fiancée, Ethel du Pont, was lovely and wealthy, the favored daughter of one of the most powerful men in America. Franklin Jr. and Ethel had been carrying on a romance ever since her coming-out party. It had been a very private courtship. Their families, one rock-ribbed Republican, the other ultraliberal Democrat, were, to put it mildly, at political odds, and neither wanted the publicity. Eventually, of course, the newspapers found out that FDR Jr. loved Ethel du Pont.

Tonight he was unhappy. It was November 20, 1936, just weeks after his father's reelection as president of the United States, and here he was at the geographic center of Rhode Island Republicanism, the Hock Popo Ski Club at the Agawam Hunt, dressed in a ridiculous outfit, in the middle of a costume ball given by some of the richest people in the nation. The hall was full of the sort of plutocrats and industrialists who thought his father's "New Deal" was something akin to a socialist revolution. Here he was, the president's son, wearing leather lederhosen, a bolero jacket, and a green Tyrolean hat with feather, Ethel in a peasant dirndl skirt, straw hat, and puffed-sleeve blouse trimmed with edelweiss—cartoon Germans at a time when his father was beginning to take on Hitler in public speeches.

The week since their engagement announcement had been stressful, a string of cocktail parties and formal dinners, starting with a glittering bash thrown by the du Ponts at Owl's Nest, the family estate in Delaware. There was endless and endlessly tiresome press attention, most often presented in the form of Romeo-and-Juliet stories featuring attractive young lovers from warring families.

Now, to top it off, he was not feeling well. Maybe he was catching a cold. He was accustomed to good health—strapping and handsome Harvard senior and crew member that he was—but in the past few days he had begun to feel run-down. His head felt like it was full of concrete. His throat was sore. Still, Ethel was having a good time, so they stayed a respectable amount of time, drinking and dancing.

Five days later, the day before Thanksgiving 1936, FDR Jr. was admitted to Massachusetts General Hospital in Boston. His admission papers listed the reason as an acute sinus infection. That was the beginning.

WHEN SHE HEARD about her son, Eleanor Roosevelt made certain, first of all, that he was installed in Phillips House, the section of Mass General that was more like an exclusive hotel than a hospital. Then she canceled Thanksgiving plans at the White House. The diagnosis did not sound dire, but no son of hers would spend a holiday in a hospital alone. President Roosevelt, on a goodwill trip to South America aboard the USS *Indianapolis,* wrote her, "Dearest Babs . . . I'm sorry about Franklin's sinus. I hope it will clear up quickly and think it will if he will go to bed early for a week." She rushed to spend the holiday in Boston, where the attending physicians assured her that she had nothing to worry about. Sinus infections generally resolved themselves in a few days. If he got his rest, it was more than likely that FDR Jr. would be released no later than the following Monday. She telegraphed the president, FRANKLIN JR MUCH BETTER THINK HE CAN LEAVE FOR SOUTH IN A FEW DAYS.

Then she flew home. There was little more she could do, and she was needed at the White House. In any case Ethel du Pont was on her way to stay with the patient. All, it seemed, would be well.

Five days later the First Lady rushed back to Boston, this time by car, through a pouring rain. Something had gone wrong. FDR Jr. was not getting better. His fever was reaching dangerously high levels. After a restless night, he seemed to rally, and the next day the hospital physicians again assured the First Lady that he was going to be fine. It was just a small setback. He would be released shortly. Again Mrs. Roosevelt returned to Washington.

But her son was not released. Publicly, a stream of recurring, reassuring public statements continued to flow from hospital administrators and appeared in newspapers on an almost daily basis. Privately, the patient's condition continued to deteriorate.

After another week with no improvement, Mrs. Roosevelt took charge. She consulted her friends in the Boston area and arranged to bring in the best ear, nose, and throat man in the Northeast, Dr. George Loring Tobey Jr., as her son's personal physician. Tobey was fifty-five years old and at the peak of his skills. He was of the wrong political persuasion—a staunch Republican with a love of expensive suits, a blueblood who could trace his lineage back well before the American Revolution—but he was also a fine physician, a "fashionable and crackerjack Boston ear, nose, and throat specialist" as a journalist of the day put it. He was well trained, well respected, well connected, and highly recommended.

After examining the president's son, Tobey ordered immediate changes in his care. He was concerned about a tender spot forming under FDR Jr.'s right cheek; Tobey thought it was likely to be a pocket of infection, a small but growing abscess, perhaps an important part of the reason the young man was not getting better. He suspected strep because of the throat but ordered lab tests to see exactly what sort of organism was causing the problem. Then he asked the First Lady's permission to perform an operation to drain it.

Suddenly things began to go seriously wrong. Before Tobey could operate, FDR Jr.'s throat began to close, making it difficult to breathe. His fever spiked to an even more dangerous level. He began coughing up blood. It appeared that he was hemorrhaging from the throat.

Tobey made some fast decisions. The lab tests, when they came back, showed that FDR Jr.'s abscess and throat infection were caused, as he had suspected, by strep. This worried Tobey deeply. The strep had gotten into FDR Jr.'s sinuses. The bleeding in the throat could mean that strep was destroying the tissue there, possibly opening a passageway to an infection of the bloodstream. A blood infection with this particular germ would make everything up to this point look mild. A strep blood infection, after all, had killed the son of another president just twelve years earlier.

The wild card was the bacterium itself. Strep was very bad, as Tobey well knew, the cause of millions of deaths. Some strains of strep, however, were far worse than others. There are factors in our favor, he told Mrs. Roosevelt. Your son is strong, young, and physically fit, and the staff at Massachusetts General is giving him the best possible care. He then advised Mrs. Roosevelt to return to Boston as quickly as possible.

In addition to his other attributes, Tobey read the medical literature voraciously, keeping up with all the latest developments. Unlike many physicians, he was not afraid to try something new. He had heard about Prontosil and had been among the first in the United States to try it on a few patients with severe strep infections. He had seen good results. He knew, too, about the spate of recent work by Long, Bliss, and the others at Hopkins. It was increasingly clear that sulfa could work in advanced cases. It was still unproven in his view, still experimental, and it went against the grain to try an experimental drug on the president's son, but he was running out of options.

He and Mrs. Roosevelt talked about Prontosil. Before she would approve its use, she wanted to hear more about it. She called Perrin Long to find out firsthand about his experience with the drug, its side effects, and how it might relate to her son's case. She hesitated about using her son, as she wrote later, as a "guinea pig," but after talking with Long and finding out that the side effects he was seeing were very mild, she gave Tobey permission to try it.

PRONTOSIL SOLUBILE was a beautiful drug. In this liquid form, it had a vibrant, rich color that physicians described as port red or ruby red; in a glass vial or syringe, it caught the light deep inside and shone. It was the color of blood, but with a clarity, a purity that real blood never had. In mid-December, during FDR Jr.'s third week in the hospital,

Tobey administered the first injections. Dosage was still a question mark; given the patient's deteriorating condition, Tobey erred on the side of aggressive therapy and dispensed generous doses, injecting the red liquid and managing to get down his patient's throat some of the white pills he had obtained from Bayer's U.S. arm, which was getting ready to sell them under the trade name Prontylin.

There was little change in the first hours. Reporters now knew there was a story here, and they began asking Tobey questions. On December 13 he told them he was waiting for FDR Jr.'s fever to lower; it was still too high to allow the operation he wanted to do for the sinuses. The press asked if the young Roosevelt would make it home for the holidays. Tobey said he was "dubious."

Both Ethel du Pont and Mrs. Roosevelt were at his side now. "We still continue to spend a great many hours at the hospital just waiting around," she wrote, "for I never knew the time when hospital routine worked out exactly as it was scheduled to work. I look back over a rather long life, and waiting around is one of the things which are inevitable where illness is concerned. So you want to keep some occupation handy which you can pick up whenever you are told politely to step out of the room, or just to wait until something or other is done. The really busy people are the nurses and the doctors, so you are glad enough to wait patiently and store up energy for the time when you may need it." Eleanor Roosevelt spent her waiting time on a chair in the corridor outside her son's room, answering piles of correspondence.

The injections were tingeing Franklin Jr.'s skin red, a sign that the drug was working, Tobey told her. His skin would soon return to its normal shade. Tobey was waking the patient to give more medicine every hour, taking blood samples and tracking his temperature. During the first long night, there was still little change. At least he was not getting worse.

Then, finally, on the fourteenth, FDR Jr.'s fever began to subside. Tobey noted that the swelling in his throat was easing. By the middle of the day, he was sleeping more easily and seemed to have more energy when he woke. The infected area in his cheek was no longer as inflamed or tender. His temperature dropped to normal in the space of a few hours. He woke up hungry. Tobey told reporters that evening that the patient was "very much improved"—so much so that an operation was no longer needed.

In fact, there was a complete turnaround. FDR Jr. felt so much better so quickly that Tobey had to enforce an unusual rule: In order to prevent damage to the healing throat tissue, the patient was forbidden to laugh.

WHAT HAPPENED WITH FDR Jr. was not as important as what happened next. A team of reporters from the *New York Times* kept after Tobey and soon got him to talk about Prontosil. They broke the story on December 17 under a front-page headline that read YOUNG ROOSEVELT SAVED BY NEW DRUG. The story caught the imagination of the nation, went out over the newswires, and was widely reprinted in both the United States and leading newspapers worldwide. During the next weeks, it made it into the newsweeklies, onto the radio, and into monthly magazine features. It was an enormous story for a short time, the first most of the public had heard about Prontosil and the other sulfa drugs. It was a positive story, a medical triumph, a welcome relief from the endless string of dark stories about the tribulations of the Depression. It was the spark that would ignite worldwide demand for the drug. After reading about the Roosevelt cure, people began writing Mass General, Tobey, FDR Jr., even the White House, asking for the drug, demanding to know more, and offering free advice. Several young women, apparently unaware of Ethel du Pont's presence, proposed marriage. One man asked Mrs. Roosevelt why FDR Jr. wasn't being cared for at the White House so that the money saved on his hospitalization could be sent to him or some other needy person. Many of the inquiries were from sufferers of chronic sinus infections; Tobey had to make public statements trying to distinguish between that condition (which could be caused by a number of different bacteria or viruses, many of which did not respond to the new drug) and strep infections. The demand for Prontosil and its derivatives in the U.S. soared.

Prontosil saved FDR Jr.'s life, but he continued to suffer from a severe sinus infection. He was so weakened that he spent not only Thanksgiving but Christmas in Phillips House. At Christmastime, Mrs. Roosevelt wrote in her daily syndicated newspaper column, "I hated to leave [Washington], but we couldn't any of us bear to think of Franklin Jr. alone by himself, so here I am in Boston." Her trip had a second purpose: She was, she wrote in a private letter, "weary and depressed" from the work of getting her husband's second term rolling and "exhausted" by the "painful and perilous" illness of her son. Going to Boston allowed her an escape from the elaborate formal rituals of the season. While FDR Jr. slept, she could read and relax.

By now, FDR Jr.'s case did not require vigilant attention from members of the family. He would eventually fight off his sinus infection and in June would marry Ethel du Pont (the first of what would eventually total five marriages). He would serve and be decorated in World War II. He would be elected to three terms as a U.S. congressman. But nothing he did would equal the impact of his illness and recovery. Linking the most famous name in America to the curative powers of Prontosil kicked off America's great sulfa boom.

CHAPTER SIXTEEN

SULFA WAS SLOW coming to America. By the end of 1936, when FDR Jr.'s case first hit the newspapers and the U.S. public was introduced to the new medicine, sulfa was already well known and becoming widely used in Germany, France, and Great Britain. Following the publication of Colebrook and Kenny's first study of the drug's astounding effect on childbed fever victims in the summer of 1936, the medical community awoke to the fact that Prontosil was more than just another overhyped German chemical cure. This was a breakthrough. After a quarter century of pessimism, chemicals were suddenly the talk of medicine again. "The whole field is one of the most fascinating that has been presented for some considerable time alike to the clinician, the chemist, and the bacteriologist," enthused a British biochemist after the childbed fever numbers were published. Prontosil in its various forms—including the pure sulfas, Prontosil album (the German version), and Septazine (the French)—was rapidly adopted by maternity hospitals and everywhere that strep diseases were treated. It was cheap, it was lifesaving, it had few side effects, so why not give it to everyone? Its use spread rapidly—to some observers, alarmingly. When Colebrook found that obstetricians were prescribing sulfa not only for childbed fever patients but as a preventive medicine given to every woman who arrived to give birth, he felt compelled to write a strong letter to *The Lancet*. In it he reminded everyone that his studies did not indicate that the drug should be given to otherwise healthy pregnant women in hopes that it might stave off a possible infection. He then noted that the drug was not without side effects—nausea, changes in blood chemistry, some disquieting effects on kidney function in rare cases, a few reports of allergic reactions—and that practitioners would see more side effects the more they used sulfa. In other words, sulfa was not to be handed out like candy.

But the stopper was already out of the genie's bottle, and there was no putting it back. Adoption of the drug accelerated rapidly in Europe during 1936 and 1937. "The enthusiasm for this group of compounds in Germany and France is nothing sort of astonishing," a British physician noted. Physicians were using it on a lengthening list of strep diseases, each of which seemed to respond, as well as on staph and urinary tract infections (where pure sulfa was giving good results). No one knew how it worked, but everyone was starting to use it.

For a short time, Bayer reaped the lion's share of profits. The company's efficient, powerful, and smooth-running marketing and sales machinery provided it with an edge over any competition in Europe. Bayer's Prontosil in its lengthening list of forms—tablets and solution, red and white—was advertised widely, the respected Bayer cross displayed over the delicate profile of a red-haired female patient (a subtle reference to the color of the company's original medicine). The virtues of the drug were outlined in detail in physicians' brochures that described its ability to fight strep as well as—in the case of white Prontosil—staph and urinary tract infections. Pure sulfa, like the original Prontosil tablets, was almost impossible to dissolve in liquids, giving Bayer a sales advantage because the firm's Prontosil solubile the only available liquid form of the drug. This was especially important in hospitals, where the sickest patients needed injections rather than pills. Sales of Prontosil rose steadily in Europe through 1936. The strategy to establish the Prontosil line with its growing number of trade-name variations (Prontalbin, Prontylin, Prontalin, Prontosil album, Prontosil rubrum, Prontosil solubile, and so forth) as the world's premier sulfa drug looked like it was working.

THE EXPERIENCE in the United States undid that strategy. It started out well for Bayer with the publicity surrounding FDR Jr.'s illness, a very widely reported story in which every article mentioned Prontosil or Prontylin (the trade name under which the company was marketing pure sulfa in the United States) by name, a windfall of news coverage equivalent to millions of dollars of free advertising. The publicity led quickly to a surge in sales that caught the company by surprise. Its U.S. subsidiary was unable for a time to meet the mushrooming demand for the miracle drug that had saved the president's son. The company's patents were already secure in the United States—when the story broke at the end of 1936, Bayer had established patents for Prontosil in America and practically everywhere else (except, of course, France)—but its U.S. subsidiary, the Winthrop Chemical Company, low on supply when the Roosevelt story went nationwide, had to delay shipments, and then, just when adequate amounts of both red and white Prontosil became available, was held up by another problem. Winthrop had provided pure sulfa to FDR Jr.'s physician under the trade name Prontylin; unfortunately, however, the name had not been approved for use in the United States by the Council on Pharmacy and Chemistry, the arm of the American Medical Association charged with testing and approving new drugs. The U.S. council would not allow sales of the drug under a trade name without the permission of its discoverer, and for the purpose of marketing Prontylin, the council deemed that the discoverer was not Klarer or Domagk or anyone at Bayer. Pure sulfa, the group concluded, had been discovered in Fourneau's laboratory in Paris. Any trade name that Bayer wanted to use in the United States would require Fourneau's permission. Distribution would be halted until permission was granted. Heinrich Hörlein was so informed.

It was humiliating. But Hörlein swallowed his pride and asked Fourneau for his permission, inserting the request in a letter he wrote to Fourneau in February 1937. Hörlein started with a litany of complaint concerning Fourneau's work in regard to Prontosil and other German drugs, noting "the intellectual damage that we [Bayer] have suffered" from Fourneau's claim to have discovered sulfa while after all they, the Germans, had been pursuing it for years. Fourneau responded in March 1937, "[T]he people in our laboratory feel—and many foreign chemists who have met or corresponded with us also feel this way—that we actually made a discovery in the immense field opened by prontosil [*sic*. Fourneau lowercased the name Prontosil in all his correspondence]. Nothing indeed, in the communications you made before ours, allowed us to think that you had yourself recognized the activity of [pure sulfa] and it would be difficult to understand, in that case, why you would not have put prontosil-album on the market right away." As for the name Prontylin, Fourneau wrote, he had given the French drug company Rhône-Poulenc all rights to discoveries made in his laboratory. Hörlein would have to ask permission of the directors of that company. "I would be very happy, naturally, to have a chat with you whenever possible, because I really want our personal relations to remain as cordial as possible," Fourneau concluded, "but I do not believe that our conversation could lead me to that paradoxical result that, alone among French chemical researchers, I should stop working in the same fields you do."

The delay in securing a trade name, added to Bayer's supply glitches, slowed distribution of Prontosil in America during a period of intense interest. American buyers began to look for sulfa elsewhere. They had no trouble finding it. U.S. drugmakers, as soon as they became aware of Fourneau's results and learned that pure sulfa was effective, unpatentable, cheap to make, and freely available for general use, quickly filled the gap. In the first six months of 1937, the major U.S. drug firms Lederle; Lilly; Merck; Squibb; and Parke, Davis all began producing and selling their own brands of sulfa. Those were just the big players. So many minor U.S. firms jumped on the sulfa bandwagon by mid-1937 that no one could keep count. By the end of the year, consumers could buy pure sulfa over the counter at their local drugstores under twenty-odd trade names. Sulfa sales soared. Prontosil was only one of many brands available.

The sulfa boom in the late 1930s was also fueled by a continuing string of positive medical stories, in both professional journals and mass media, about sulfa's miraculous effects. In November 1936, just before the FDR case, Long and Bliss delivered the first American medical report of large-scale success with the new drug, telling a public meeting of U.S. physicians the results of their experience with seventy strep patients at Hopkins who suffered from everything from skin rashes to massive abdominal infections. Many of the cases had been quite advanced. Much interest focused on children with meningitis, one of the deadliest childhood diseases. Meningitis resulted when bacteria invaded the spinal fluid, turning it milky. It could be caused by a number of different bacteria, the most common of which was meningococcus, a bacterium that sulfa so far had been powerless to fight. But strep could also cause meningitis. When it did, it was a death sentence: In the days before sulfa, only two children out of every hundred with strep meningitis survived. Using sulfa, Long and Bliss reported, they had saved nine out of ten cases at Hopkins. Typical was a five-year-old girl who presented on December 7, 1936, with a temperature of 106 degrees, jerky movements of the eyes (indicating that the muscles around her eyes were paralyzed), and thick, cloudy spinal fluid. She was a few hours away from death. They treated her aggressively with sulfa. Within a few days, she was sitting up drinking juice. On Christmas Day she was sent home, cured.

"Drama? You bet," enthused a typical sulfa story in *Collier's,* one of the nation's most popular magazines, in the spring of 1937, the year the sulfa boom hit America. The media loved mixing selfless young medical researchers like Long and Bliss with children snatched from certain death. Long and Bliss's work was touted everywhere from *Fortune* to the *New York Times,* along with Colebrook's, Domagk's, and Fourneau's. The stories were so common that by August, *Time* magazine headlined yet another article AGAIN, SULFANILAMIDE. Long and Bliss began receiving scores of letters from suffering patients and curious doctors, asking about uses, doses, side effects, and methods of delivery. They were soon spending so much time dealing with sulfa correspondence and calls that they had little time for anything else. They could not wait for the clamor to end so they could get back to work.

AT HÖRLEIN'S ELBERFELD factory in Germany, the drug development group pulled themselves back together and began to function once again as a team. There was plenty to be happy about, despite the revelations regarding pure sulfa. Domagk, Mietzsch, and Klarer were jointly awarded one of Germany's highest chemical honors, the Emil Fischer Memorial Medal, from the Association of German Chemists. In public statements the three men papered over their conflicts, with Domagk commenting, "I can see that further successes for the cooperation between physicians and chemists concerning chemotherapy of bacterial infections are in store. Just like husband and wife are dependent upon each other in a good marriage, so are physicians and chemists when it comes to research in the area of chemotherapy even if certain tensions will be inevitable."

It was clear, however, who Domagk considered the husband and who the wife when it came to getting credit for Prontosil. The news stories and journal articles around the world focused primarily on Domagk and his dramatic Christmastime 1932 experiments with mice, mentioning the chemists as supporting players, if at all. Public attention converged on Domagk in part because only his name appeared as author of articles in medical journals, giving readers the idea that only he was central to the research, and in part because of general ignorance about the cooperative nature of large industrial research laboratories. Mietzsch and especially Klarer might feel unjustly neglected as Domagk received letter after letter congratulating him for his lifesaving work, but Hörlein kept them focused on more important things.

With pure sulfa now replacing Prontosil as the medicine of choice in the United States, and with an increasing number of drugmakers beginning to make their own versions in Europe and Asia as well, it was vital that Bayer move on to find next-generation sulfa drugs, improved versions that would work better on more types of diseases.

Domagk, Klarer, Mietzsch, and the new chemist on the Bayer team, Robert Behnisch, threw themselves into the search. As the senior man in the lab, Mietzsch was put in charge of organizing the chemists. He approached the mission logically. The sulfanilamide molecule consisted of a single central carbon ring with a sulfur-containing chain on one side, directly opposite a nitrogen-containing chain on the other. It appeared from what they already knew that the molecule could be modified at those two places, the side chains on opposite sides of the central carbon ring, without losing its power. He assigned Klarer the job of creating variations at one end and Behnisch the other. Mietzsch himself focused on seeing if anything else could be done with the ring at the center. He quickly found that any change in the ring greatly lessened or eliminated the potency of the drug (this was the problem they'd had in 1934, when a three-atom addition to the ring had rendered pure sulfa useless). The chemists began getting the best results at Behnisch's end by linking together two sulfa molecules end to end. In 1937, Domagk found that these double-sulfa molecules were effective against not only strep but also gonorrhea, staph, and—surprisingly and gratifyingly—his old nemesis, gas gangrene. After quick testing by physicians, Bayer launched one of these new-generation sulfa medicines in 1937 under the trade name Uliron. Research continued, and an even more effective variation, Neo-Uliron, was found and marketed two years later. By then, however, Germany was at war. While both Uliron and Neo-Uliron were widely used in Germany, neither achieved sizable worldwide success.

Domagk kept up a steady stream of publications on the sulfa drugs, confirming his growing stature as the man most responsible for the discovery. His fame was recognized both at home and around the world. In 1937, he was invited to become director of the pathology section of the Kaiser Wilhelm Institute of Medical Research in Heidelberg, a very great honor. The answer to the KWI came not from Domagk but from Hörlein, who politely rejected the offer, noting that Domagk was too busy following up on a number of imminent sulfa breakthroughs and was needed at Bayer to "let the apparatus created by us work as quietly as possible."

QUIET WORK on sulfa drugs was almost impossible. Publicity and controversy seemed to follow the medicine. Fourneau and Hörlein were still feuding, the French chemist now expressing his disbelief that the Germans could have researched the Prontosil molecule for more than two years without finding that its power came from pure sulfa. Instead he assumed that Hörlein and the Bayer team had discovered the medicinal properties of pure sulfa during that long quiet spell between the December 1932 patent and Domagk's February 1935 publication, and that the company had suppressed the discovery in order to increase profits. In a way it was a sign of Fourneau's respect for German chemistry: He took it for granted that the Germans could not have missed a discovery so basic, in hindsight so logical, that his team in Paris had done it within a couple of months. He seems to have ignored the fact that his own research team stumbled across the answer quite by accident, thanks to Bovet's four extra test mice.

Perhaps Fourneau's pique was also due to Hörlein's studied approach to ignoring the French contribution to sulfa science. It was not until 1937, two years after the French finding, that Hörlein could bring himself to mention Fourneau's name publicly, and even then he downplayed the Pasteur Institute's contributions, telling a British audience finally that "Ernest Fourneau's collaborators" had been "the first to make public" the fact that pure sulfa had the same effect on streptococcal infections as did his firm's azo dye. Even Bayer's marketing materials studiously avoided giving the French any credit, with a Bayer marketing flyer of the period going so far as to relocate the discovery of pure sulfa from Paris to Elberfeld: "After the successful discovery in our laboratories of 'Prontosil' solubile (ampoules) and 'Prontosil' tablets, . . . we continued research along the same lines, basing our work on the active principles involved, and found several colorless therapeutic substances of a somewhat similar nature."

Fourneau had good reason to be irritated. His accusations that the Germans had known about pure sulfa and then hidden the discovery struck a responsive chord with other researchers, especially in Britain, and soon it became more or less accepted that Bayer had suppressed news of a lifesaving medicine in favor of profits. Hörlein's reply in a letter to Fourneau stressed the "heavy series of disappointments" he had undergone with Prontosil and complained in turn about the "misunderstanding" the subsequent French work had caused. Of course, he wrote, Bayer had been looking for its own colorless derivatives. "In fact our labs were already more advanced in that field than you were. . . . It is only because we wanted to support the lab results in the most exact way through clinical experiments, just like we did for Prontosil, that we could not publish our findings at that very time." It is a surprising statement. No available correspondence from the period in the Bayer Archives indicates that Bayer was performing clinical tests on any substances like plain sulfanilamide, and none of Domagk's lab notes or the chemists' monthly reports point in that direction. It is likely that Hörlein was simply trying to reclaim credit for a French discovery that had embarrassed his firm and him personally.

"I understand very well why you felt disappointments," Fourneau replied. "I have a harder time understanding why you reproach me." He concluded, "It is better to get along than to fight." Fourneau reminded Hörlein that the French team had always been careful to credit the work done at Bayer. He left unspoken the fact that the Germans had not returned the favor.

He would have appreciated the credit. Through the end of 1936 and into 1937, Fourneau was at odds with the administration of the Pasteur Institute, who were, he thought, less than supportive of his efforts to make an even greater triumph of the sulfanilamide discovery, missing a chance to take a leading role in the development of a fast-growing field. Everyone was starting to make improvements on pure sulfa. The Pasteur Institute was about to be left behind.

AN IMPORTANT PART of the reason America lagged a year or two behind Europe in adopting the sulfa drugs was a lack of coverage in U.S. medical journals. Before the FDR Jr. case, through 1935 and 1936, as Prontosil was beginning to achieve wide success in Europe, the *Journal of the American Medical Association* did not feel compelled to mention a word about Prontosil or sulfa—not even a note from the journal's Berlin correspondent. The word "sulfanilamide" did not appear in the *New England Journal of Medicine* until November 1937. It was a reflection of the inherent medical conservatism of the day.

Physicians had seen too many announcements of too many miracle medicines. Claims for wonder cures were the stuff of medicine-show hucksters and patent-medicine advertisements. The news coming out of Europe was interesting but incredible. It seemed too good to be true. So the medical editors chose to ignore it.

By late 1937, however, U.S. public interest in the drugs was so intense, sales and use increasing so rapidly, that the medical establishment was forced to pay attention. There was more science directed toward the new drugs as well. Research on the drugs' metabolism, fate in the body, and mode of action—studies that had begun at Hopkins in the United States—began spreading to other universities and commercial pharmaceutical laboratories. Physicians began extending their use to other diseases. U.S. researchers confirmed the European success with erysipelas—St. Anthony's Fire—against which the new sulfa drugs worked brilliantly. In 90 percent of the cases, the spread of the burning rash was halted within twenty-four hours. The death rate from erysipelas was cut by nearly two-thirds. Childbed fever mortality rates were cut even more. Successes were achieved in treating scarlet fever, pyelitis, meningitis, gas gangrene, cellulitis, otitis media, and tonsillitis. Perhaps more important, Prontosil and the other sulfa formulations were found to work against both strep-caused pneumonia and, in at least some cases, against pneumonia caused by another related germ, pneumococcus. Pneumonia was a major killer. And sulfa was showing an effect against an even more common (and embarrassing) disease. Gonorrhea was a worldwide epidemic problem, rarely fatal but always problematic, cause of many visits to the doctor and much lost work time. Nothing had ever been very effective as a treatment. Pure sulfa did not work perfectly, either, but it worked sometimes. While researchers looked for better variations on the drug, the public got wind that there was a cure for the clap, and demand for sulfa increased even more.

During 1936 in Europe and 1937 in the United States, hundreds of chemists and dozens of pathologists at scores of firms started running sulfa tests on thousands of mice. Pure sulfa could not be patented, but if it was attached to a new structure, the resulting compound could be. Scores of patents for new sulfa variations were quickly filed, then hundreds, then thousands. Some of the manufacturers paid as much attention to safety and quality as Bayer did. Many did not.

The Bayer team looked on in despair. At first they kept careful track of the new sulfa preparations competing with the Prontosil line but soon gave up. There were simply too many. In May 1937, Domagk wrote a fellow physician that he was "astonished in this case how quickly the competition brings out such preparations." The list by then included Rubiazol, Rubidium, Ectiazol, and Cellubiazol in France; Pratanol in Japan; Diseptyl in China; Streptopan in Holland; Stopton in Brazil; Supron in Prague; Streptazol in Zagreb; Ambecid in Budapest; Antiseptin in Warsaw; Streptocide, Colsulanyde, Sulfamidyl, Stramid, Sulphonamid P., and a host of others in the United States. One historian estimated that thirty-six different sulfa-containing preparations were available in London within four months of Colebrook and Kenny's first publication. Drugmakers took advantage not only of sulfa's easy availability but of Bayer's slow and careful response: While the giant German firm patiently tried to build Prontosil into a dominant brand, a hundred smaller firms jumped in with snappier names (taking advantage of the fact that Prontosil solubile, as one observer noted, "is more like a speech than a word"), less restrained advertising claims, and next to no safety studies. New sulfa preparations, each slightly different from the others, were made faster than pathologists could test them or physicians could be educated in their proper use. And they sold. Pure sulfa by itself could cure someone of a potentially fatal disease at a cost of about thirty cents a day, compared to common patent medicines that cost as much as six dollars a day and did little if anything.

The year 1937 was the start of a sulfa gold rush, and everyone went prospecting. Patients demanded the drug because news stories hailed sulfa as "the most sensationally valuable new medicine in many years" (*New York Times*), "one of the most spectacular medical triumphs of our generation" (*Science Digest*), and "a modern miracle" (*Collier's*). "I remember the astonishment when the first cases of pneumococcal and streptococcal septicemia were treated in Boston in 1937," recalled physician Lewis Thomas. "The phenomenon was almost beyond belief."

Enthusiastic physicians began prescribing it for diseases against which it had been proved effective, and many times for diseases for which it had never been shown to do anything, including conditions as mild (and completely unresponsive to sulfa) as colds and the flu, dosing any patient who presented with a fever of unknown origin. By all accounts it was relatively safe, it might lead to a cure, and it might help fight off a subsequent, perhaps more serious, infection, and in any case prescribing a few tablets was always easier—and sometimes more effective—than completing a complicated diagnosis. Nurses, too, began giving it to patients if they felt it might do some good. Future U.S. surgeon general C. Everett Koop, a medical student in the late 1930s, remembered hospitals doling it out "like aspirin." Veterinarians began using it. Dentists began using it. Demand grew so quickly that even without coming up with a new chemical variation, a drugmaker could repackage pure sulfa, jack up the price, and make money. By October 1937, Perrin Long estimated that the top ten U.S. firms alone were making more than ten tons of sulfa drugs per week. Even that was nowhere near enough to meet demand.

Given the sudden surge in use, the editors of the *Journal of the American Medical Association* felt compelled to break their long silence on sulfa in the fall of 1937, warning readers that the rapidly proliferating new forms of the medicine might prove more toxic than earlier sulfa drugs. The next month the *New England Journal of Medicine* cautioned physicians that much remained to be learned about dosage, noting the dangers of doctors' setting their own doses on a case-by-case basis. Journals for physicians dealing with children, however, where the most consistent, most striking results from the new drugs were being seen, were more enthusiastic. The *Journal of Pediatrics* devoted its entire August 1937 issue to sulfa. "No therapeutic agent has appeared on the medical horizon for many years which has attracted so much attention and interest as sulfanilamide," the editors wrote in a piece introducing the issue. "The few brief published preliminary reports of its use are literally amazing." They also noted that the drug was being used "widely and indiscriminately in private practice."

Everyone was taking it, but no one really knew much about it. No one knew if it was safe to take over a long period of time. No one in 1937 had ever done the sorts of double-blind, randomized, long-term studies on patients that are routinely done today. No one had ever seen a drug like sulfa. But it didn't matter. There was no holding it back.

THE 1937 Paris Exposition was a fantasyland of international cooperation and progress. Billed as a celebration of art and science, the exposition, mounted on seventy acres in the center of the city, featured captivating and educational pavilions from many nations, motorboat races on the Seine, "Visions of Fairyland" for the children, music and dance recitals, horse races, wine tastings, flower exhibitions, and boxing matches. Every night the Eiffel Tower and the fountains of the Trocadero were brightly illuminated. More than 34 million visitors attended the exposition during its summer run.

Particular attention was paid to the special shows and exhibits, the New Museum of Modern Art, the Palace of Light, the Palace of Refrigeration, and—striking a discordant note among all the displays of international cooperation—a large mural in the Spanish Pavilion depicting a bombing of a Spanish city, called *Guernica,* created by a painter named Pablo Picasso, and generally reviled by those hoping for a more peaceful view of international relations. The bombing was linked to the Luftwaffe, and the German fair guide dismissed it as "a hodgepodge of body parts that any four-year-old could have painted." The exposition after all was supposed to be about hope for the future.

The Spanish Pavilion sat in the shadow of what was by far the largest of the national exhibits at the fair, the German Pavilion. Clean and modern in design, it was built to impress, featuring a five-hundred-foot-high stone tower capped with a giant eagle clutching in its claws a shining gilt swastika, a symbol of Nazi power lit at night with floodlights. Albert Speer, its designer, was charged with making certain that Germany's building overwhelmed any other, especially the USSR's. He succeeded. The German Pavilion was a monument to political propaganda both outside and in, where it contained a carefully chosen mix of state-sanctioned art and examples of advanced German technology. Among the artistic contributions was a stunning documentary by a young woman named Leni Riefenstahl. Her film, *Triumph of the Will,* was honored with one of the exposition's gold medals.

An unexpectedly popular draw at the exposition was a relatively small hall hidden away behind the Grand Palais. The Palace of Discovery, as it was called, attracted more than 2 million visitors, five times the number that visited the modern art exhibit. They came for wonder and hope. The wonder was provided by exhibits including a huge electrostatic generator, like something from Dr. Frankenstein's lab, two enormous metal spheres thirteen feet apart, across which a 5-million-volt current threw a hissing, crackling bolt of electricity. The hope came from the very nature of science itself. Designed by a group of liberal French researchers, the Palace of Discovery was intended to be more a "people's university" than a stuffy museum, a place to hear inspiring lectures on the latest wonders of science, messages about technological confidence and progress for the peoples of the world. The mastermind behind the Palace of Discovery, French Nobel Prize laureate Jean Perrin, wrote, "In spite of the wars and the revolutions, in spite of the economic crisis and unemployment, through our worries and anxieties, but also through our hopes, civilization's progress is going faster and faster, thanks to ever-more flexible and efficient techniques, to farther- and farther-reaching lengths. . . . Almost all of them have appeared in less than a century, and have developed or applied inventions now known by all, which seem to have fulfilled or even passed the desires expressed in our old fairy tales."

The hope of science was one of the exposition's major themes, and German scientists took full advantage. Ernest Fourneau helped to ease their way, securing funds from the French government specifically for encouraging Franco-German scientific cooperation. German discoveries were prominently featured in the Palace of Discovery as well as in the German Pavilion itself. They included the first public exhibits of the new "synthetic bactericides"—sulfa drugs—discovered in the Bayer laboratories.

Domagk's work, too, was recognized with one of the Paris Exposition's gold medals.

III

The Left Side

Science is one thing, wisdom is another.

Sir Arthur Eddington

CHAPTER SEVENTEEN

DR. JAMES STEPHENSON wanted to know what the hell was going on. A longtime Oklahoma country doctor, Stephenson had seen plenty of death. But something did not sound right about the rash of fatalities that started hitting the people in his town in the fall of 1937.

They were children, most of them, and no physician likes to lose a child. Toward the end of September, a dozen children and a few adults had started showing up at Tulsa doctors' offices with severe abdominal pain. Nothing seemed to stop it, not even opiates. Then they stopped urinating. They fell into stupors—sometimes there were convulsions—then comas. Within a few days, six of them had died.

Stephenson and his colleagues were thinking kidney failure. That would explain the lack of urination and fit some of the other symptoms, too. Perhaps some infectious agent, a bacterium or virus, was attacking the kidneys. But the cases were spread all over the county. The patients had no contact with each other, and the people around them were not coming down sick. There was no pattern—that was the part he did not like. The patients were mostly young, but not all of them were. They were both sexes, some in town, some in the country. No sign of contagion. It could be poison. But either a poison in the water would have affected a lot more people or the deaths would have concentrated around a single source like a well or a stream.

Six unexplained deaths in a town the size of Tulsa was a serious matter in 1937, and Stephenson, head that year of the local medical society, was a serious doctor. He alerted the other physicians in the area about the strange malady, then started poring over his books and journals looking for answers.

The newest issue of the *Journal of the American Medical Association* (*JAMA*) gave him his first clue. Ever since the news of the cure of Roosevelt's son had hit the papers a year earlier, *JAMA*, like most medical journals in the country, had started running reports and editorials about the new sulfa wonder drugs, highlighting the drugs' brilliant success in treating meningitis and childbed fever as well as a slew of other strep diseases. Even in Tulsa, patients had begun asking for the drug, and physicians had begun prescribing it. Drug companies were jumping on the bandwagon: Sulfa pills, capsules, injectable solutions, and powders had recently hit the market from dozens of pharmaceutical manufacturers. Every drugstore in town was selling sulfa to anybody who asked for it.

What caught Stephenson's eye in the October 2 *JAMA*, however, was a new editorial that advised caution. Alarmed by the lightning speed with which the new drugs were reaching a large population of patients, the journal's editors reminded physicians that no new drug, including sulfa, was entirely without dangers. There were a growing number of reports of sulfa side effects, the editorial pointed out, ranging from increased sensitivity to sunlight to changes in blood chemistry—nothing too serious, but, "As with any new drug," they concluded, "physicians should moderate their initial enthusiasm with the proper dose of watchfulness."

When Stephenson read that, something clicked. He remembered hearing from physicians in Tulsa that several patients with the mysterious kidney problem had been taking sulfa for sore throats. He immediately checked to see what specific medicines they had been taking. And he found his suspect. The one common thread in every case was a proprietary drug called Elixir Sulfanilamide, produced by the Massengill Company.

On October 11, Stephenson wired the American Medical Association offices in Chicago, describing the mysterious deaths and suggesting that there might be a link to the elixir. Association officials wired him back: They would look into it, they wrote, and asked him to rush them a sample of the suspected medicine for testing. The AMA also wired the elixir's manufacturer in Bristol, Tennessee, asking Massengill to share the formula with them.

Dr. Samuel Evans Massengill, the silver-haired, clean-shaven, strong-jawed founder, owner, and for forty years president of the Massengill Company, was happy to oblige. He was himself a physician, although he had never practiced. He had instead built a sizable drug-manufacturing operation with warehouses in New York and San Francisco, a plant in Kansas City, and a solid reputation for offering reliable treatments to the public and the medical profession. He was the descendant of a pioneer Tennessee family and a proud southerner. His grandfather had been a country doctor who rode with Bedford Forrest's cavalry. His uncle had helped to start a medical college. The public and press might still call what his company produced "patent medicines," but he considered himself a respected maker of proprietary medicines and a member in good standing of a group of businessmen who were also proprietary medicine manufacturers. He was no huckster with a traveling medicine show.

He called in his head chemist, Harold Watkins, the man who formulated the elixir, and asked him to bring the recipe with him. It was fairly straightforward: fifty-eight pounds of pure sulfa powder dissolved in sixty gallons of diethylene glycol, some water, a handful of flavorings and colorings. Watkins had chosen raspberry for the basic flavor and put in a pound of saccharin to sweeten it. Sulfa was notoriously hard to dissolve in water, but there was a demand for a sweetened liquid form of the drug, especially for treating Negroes and children, who every patent medicine maker "knew" preferred their medicines in a sweet syrup. Everyone was making sulfa drugs, but this was a part of the market that so far had gone begging. Massengill salesmen knew they could sell a sulfa elixir. Watkins made it, and he made it taste good.

Both Watkins and Massengill were sure that their medicine could not possibly have caused any deaths. The only problem Watkins had when making it was a difficulty in getting the sulfa to dissolve. Alcohol did not work well, nor did any of the more common medicinal solvents. The only thing that seemed to work was diethylene glycol, an industrial solvent, a common ingredient in lotions and salves. That had worked extremely well. The men discussed the fact that there was always a chance of contaminants—some poison, arsenic, or heavy metals that might have accidentally gotten in during the making of large batches—but that sort of contamination had never happened before at the company. The two men resolved to look into it immediately.

Then Massengill wired Watkins's formula to the AMA in Chicago. It was an unusual step. Makers of proprietary medicines jealously guarded the recipes for their remedies. Formulas made public could become formulas copied, with profits diminished. Massengill appended a note to the telegram urging the association to keep the elixir recipe strictly confidential. He noted that his company had performed no toxicity tests on the elixir before sending it out. He added that he thought the Tulsa deaths might have been triggered by a drug mix; perhaps his elixir had been given to patients who were also receiving other drugs.

In those days drugmakers did not have to worry much about the law. The only agency charged with bringing the breakers of drug laws into compliance was a small office within the Department of Agriculture, first called the Bureau of Chemistry (because the chemists had the equipment needed to test for toxins in foods), then the Food, Drug and Insecticide Administration, then simply the Food and Drug Administration (FDA). In 1937, the FDA could field fewer than 250 agents nationwide, a tiny force compared to the mushrooming number of foods and medicines being sold in the United States. The agents spent most of their time tracking down and testing shipments of food. When it came to drugs, they could do little more than a bit of policing after the fact, testing suspect medicines after they were already on the market to make sure their labels were accurate, that they met pharmaceutical standards (if any existed for a particular medicine), and that they were not contaminated.

The FDA was not seen as much of a factor when it came to drug safety. Into the regulatory vacuum stepped the AMA, establishing its own chemistry labs to test drugs in 1906, then publishing its own list of medicines approved for professional use. Only those with an AMA stamp of approval could be advertised in medical journals. The insignificance of the FDA was underscored when Stephenson wired not the government but the American Medical Association for help. When it came to a potentially dangerous drug, it never crossed his mind to contact the FDA. The American Medical Association did not notify the FDA either. Why bother?

IT WAS NOT until October 14, three days after Stephenson first wired the American Medican Association about his problem, that news of the deaths in Tulsa finally reached the FDA offices in Washington, D.C. The agency heard about it from a physician who had heard rumors of the Tulsa poisonings circulating through the professional grapevine. After it finally got to the FDA, the news was taken directly to agency head Walter Campbell.

Two thoughts immediately occurred to Campbell. The first was how he should start investigating and stopping the deaths. The second was how this incident would play politically. Campbell had been fighting for years to make the public safer, and he believed that a stronger FDA was the way to do it. He was a scrapper from Kentucky who had spent his entire professional life in the agency, serving almost thirty years, running it efficiently, earning the respect of his staff, and constantly agitating for a bigger role. Dark, intense, balding, and indefatigable, Campbell took personal control of the Tulsa investigation.

He immediately ordered an agent to Oklahoma to talk with Stephenson and two more to the Massengill headquarters in Tennessee. FDA agent William Hartigan arrived in Tulsa on October 15, after traveling from the nearest federal office in Oklahoma City. Hartigan interviewed Stephenson, then walked the town's major streets and asked questions. Later that day he wired Campbell in Washington: six suspicious deaths confirmed in the Tulsa area, more suspected cases in hospital, elixir apparently involved, samples sent to AMA for testing, more later.

Stephenson, however, did not need any federal agent to tell him that the elixir was to blame. Before Hartigan arrived, before the AMA finished its tests, he had already told every doctor around Tulsa what he knew; created an impromptu phone tree of doctors, pharmacists, and the local hospital staff to track cases as they arose; and asked the Tulsa area Retail Druggists Association to stop selling the elixir. Now he was looking for a way to reverse the kidney damage. More patients were coming down with whatever it was that was causing their kidneys to cease functioning. But try as he might, he and the other physicians in town could not find a way to stop it.

The same day that Hartigan arrived in Tulsa, an FDA agent and an FDA chemist met with Dr. Massengill and chief chemist Harold Watkins at the company's impressive, four-story office in Bristol. They reviewed the process through which the elixir had been developed. Watkins explained that he had used commonly available ingredients to make a new product that had been requested by the public. When Watkins was satisfied with his recipe, he had sent a sample to the Massengill "Control Laboratory," where it had been tested and approved for flavor, appearance, and consistency. The formula was then sent to the Massengill plant in Kansas City, where forty gallons of Elixir Sulfanilamide was mixed and bottled. The first shipments were sent to customers on September 4, 1937. The new form of sulfa had been well received, and the company quickly produced and distributed another two hundred gallons.

This meant that by the time Stephenson first wired the AMA from Tulsa, the drug had been on the market for a month, flowing from manufacturer to distributor, to salesmen, to drugstores, to physicians, to patients. The elixir was in the hands of who knew how many people scattered around the country. If there was a problem with this medicine, the FDA representatives realized, it was going to be a very big problem.

Massengill was not upset. It was true, he said, that the company had performed no tests on animals or humans before releasing the elixir. Nor had they tested the drug to see what might happen to it once it was in the digestive system. But as the inspectors knew, companies almost never did that for any drug. Massengill assured the investigators that he was certain there was nothing wrong with either the company's formula or its preparation techniques. Perhaps, he suggested, it was the sulfa itself that was dangerous. It was, after all, a relatively new and unknown German drug. Perhaps it was interacting with something else the patients had been taking.

The FDA men were not so sanguine. They asked headquarters to check the backgrounds of Watkins, Massengill, and the Massengill Company. Watkins, the wire came back, had earned his pharmacy degree at the University of Michigan in 1901, worked for a variety of drugmakers, started working for Massengill less than two years earlier, and had been cited for mail fraud in 1929, while marketing a weight-loss potion. A check of the FDA files showed that Massengill himself had never broken the law, but his company had been cited twice for infractions of the law, both incidents related to mislabeling. The company had quickly fixed the problems. It was a relatively clean record for a proprietary medicine maker with such a long business history.

The agents asked Massengill to immediately stop all elixir sales and warn customers and salesmen of the potential danger. They could not demand that he do so under the law, they understood that the elixir had not yet been proved to cause anyone's death, but they asked him to do it for the good of the public while they sorted through the issue. Massengill cooperated, sending thousands of wires over the next few days, telling customers that they could return all stocks at the manufacturer's expense. But his telegrams contained no mention of toxicity or danger, and they were sent only to the customers Massengill had on record. The addresses of more than six hundred shipments of the elixir, in amounts from a pint to a gallon, could be tracked from Bristol. Another two hundred small samples had been sent to the Massengill sales force. By now, however, the drug had percolated through the system far beyond the original buyers, through wholesalers or retailers who resold the elixir, or physicians who distributed it to patients. The Massengill company had no way of tracking the medicine any further.

On October 16, Hartigan wired again to report that there were now nine dead in Tulsa, eight children (who had taken the elixir for sore throats) and one adult (gonorrhea). By speaking with most of the doctors in the city and visiting every drugstore, he had uncovered a total of thirty prescriptions filled before sales of the elixir were stopped. He also found time to go to a local hospital and attend the autopsy of a suspected elixir case. The coroner explained what he was seeing: Most viscera normal except the kidneys, which were enlarged, purplish, and clotted. Something was very wrong there. The cause of death was listed as renal (kidney) failure. The victim being examined was eight years old.

Hartigan also spoke with family members. They told him how it went: A few hours after taking the elixir, the victims began to feel nauseous and started vomiting. That was good, actually, because some of the patients felt so bad they stopped taking the medicine. It probably saved some lives. The unlucky ones, the ones who had taken too much or kept taking it, stopped urinating. Then, quickly, came the pain, excruciating, unrelenting pain. The mother of a six-year-old girl who died after taking the elixir wrote later, "Even the memory of her is filled with sorrow for we can see her little body tossing to and fro, and hear that little voice screaming with pain and it seems as though it would drive me insane." Some victims took two days to die, some as many as seven. Some children died after taking a mere two fluid ounces of the medicine, while some adults survived after taking as much as seven fluid ounces.

FDA chief Campbell knew all he needed to know. He wanted to immediately seize all stocks of the elixir nationwide—but the law did not allow him to. Proprietary medicine manufacturers and their lobbyists had for decades effectively derailed any regulations that would allow mass seizures of their products: If one particular batch of a drug was somehow tainted, they did not want the entire national supply seized, the entire reputation of the product threatened. Campbell was left with no straightforward way to fight what appeared to be a national disaster in the making. So he found a technicality. Massengill had called the medicine an "elixir," a term that, at least in the precise language of the agency, applied only to alcoholic solutions. But Massengill's elixir contained no alcohol. The solvent was diethylene glycol—Campbell's people were checking on that ingredient as well—which meant it was not technically an elixir, which meant that it was mislabeled. The law in 1937 allowed drugs to circulate if they were potential poisons—but not if they were mislabeled.

On October 17, the FDA wired every one of its agents, telling them to search out and seize all stocks of Elixir Sulfanilamide, Massengill.

On October 18, the FDA agents in Bristol talked again with Massengill and Watkins, telling them the latest news from Tulsa, noting the extreme urgency of the situation, pointing to diethylene glycol as a possibly toxic compound, and asking for more help in tracking down and halting the sale of the elixir. Watkins began changing his story. Now he told the agents that he had indeed done animal toxicity tests on guinea pigs before releasing the product. He could, however, produce no records. Watkins knew that the FDA was looking at the ingredient diethylene glycol, and he told the agents that during the past few days he had himself taken "teaspoonful" doses of the elixir and also drunk straight diethylene glycol. He was obviously desperate to prove that he had not inadvertently poisoned people. He told the agents that he felt fine. The agents asked that Massengill send out another set of wires, stressing more strongly the deadly threat posed by the elixir. Massengill said he would do all he could. But another day would go by—a day of continued toxicity testing by the American Medical Association, a day of waiting for the results of animal tests, a day of patients buying more elixir—before he sent out any message implying that his product was dangerous to health.

By then the death toll was thirteen, the latest four in a new city, the black section of East St. Louis, Missouri. That day in Chicago, the editor of the *Journal of the American Medical Association,* Morris Fishbein, held a press conference at which he prereleased an editorial from the next issue, warning the nation "for the protection of the public" that the Elixir Sulfanilamide was dangerous and striking a blow against patent medicines in general, noting that "this tragic experience should be a final warning to physicians relative to the prescribing and administration of semi-secret, unstandardized preparations."

Campbell put every FDA agent he had on the job. It became the largest single operation in agency history. Most of them tracked the progression of orders, from producer to the warehouses of chain drugstores and regional distributors, to drugstores and physicians, to as many patients as possible. The task seemed overwhelming. Each thread unraveled into a dozen more. At one large drugstore alone, agents had to thumb through more than twenty thousand receipts.

On October 18, the AMA finished its tests on the elixir and sent the results to Stephenson in Tulsa: The elixir sample he sent contained no toxic contaminants. The cause of the poisoning in their opinion was the medicine itself, apparently due to the toxicity of the solvent, diethylene glycol. Although it it had been used in external applications for decades, there was not a lot of medical information about the effects of the chemical inside the body. Deep in the medical journals, however, AMA researchers found three articles indicating that the liquid, taken internally, could be toxic in rats, mice, and humans. That same day the association issued a general warning to the nation to avoid taking Elixir Sulfanilamide.

The resulting panic threatened to swamp the already overburdened FDA. Every agency office was flooded with telegrams and phone calls, tips, and questions. Agents scrambled to keep up. One nine-year-old boy, worried about what he'd read in the papers, showed up at an FDA office with a bottle he had taken from his mother's medicine chest. The bottle was labeled "Elixir Iron." He wanted to make sure his mother was safe. The agents told him he had nothing to worry about—to the boy's "profound relief," the agents noted—and sent him home. Every visit, every call, every telegram took time. The agency followed up each credible lead, testing a number of other elixirs for diethylene glycol content (and not finding any). Now it seemed that every hour was bringing news of a new death.

The elixir was in too many hands. East St. Louis was a problem. As many as fifty prescriptions had been filled there before sales were halted. Almost all were for "colored people," and, in a pattern they would see again and again, agents found that many of the purchase and prescription records were almost useless, reading "Betty Jane, 9 mos. old" or "Mrs. Jackson (no addr.)," or providing no name at all. It was almost impossible to track down the people who had bought the tainted medicine.

In any case, finding the buyer did not guarantee that the problem was solved. One East St. Louis woman who bought elixir for her child had thrown it out, she told agents, the moment she heard that it was dangerous. "What did you do with it?" the agents asked. She pointed to the window. Outside in the alley, they found an unbroken bottle of elixir, containing enough sweet, raspberry-flavored liquid to kill a child.

The farther they tracked the elixir, the harder things seemed to get. Most shipments had been made to states in the South; much of the elixir went to rural areas, where people were poor and record keeping was haphazard. A significant portion of it was sold over the counter, leaving no prescription record to track. Some was sold to people concerned about venereal disease, who gave false names. Some druggists and physicians, worried about legal repercussions, were unwilling to release names. Some prescriptions were "lost," destroyed, or altered.

In South Carolina a physician claimed that he had written five prescriptions for the elixir but told agents that all the patients were fine. He would not provide names. When the FDA investigated further, they discovered that he had actually prescribed the drug to seven patients. Four of them were dead. One of the victims, a black lumbermill employee, was found only after a neighbor who recognized the symptoms pointed an FDA agent to the house of the victim's sister, who remembered her brother taking some sort of elixir before he died. She then led the agent a mile and a half through the fields, to a small cemetery on the side of a wooded hill where, on a pile of fresh dirt, the agent found dishes, spoons, and several bottles—it was the local custom to place on the grave anything that had to do with the final illness—one of which still contained about a teaspoon of elixir.

In Georgia a druggist claimed he had distributed six fluid ounces of the elixir. Agents secured the rest of the gallon he had in stock and sent it back to headquarters, where FDA chemists measured every drop to make sure that no elixir remained unaccounted for. The bottle from Georgia was missing twelve fluid ounces, not six. Confronted with the fact, the druggist admitted that he had made two additional sales. Both the purchasers were dead.

Massengill salesmen presented a different kind of problem. Most of the more than two hundred company representatives carrying and giving out samples of the elixir moved from town to town quickly, lived at hotels, and were difficult to locate. In one case the FDA picked up the trail of a salesman in Washington, D.C., followed it to Jackson, Michigan, then to Baltimore, where agents discovered that they were following the wrong man. They finally found him in a hotel in University Park, Maryland, and seized his remaining elixir samples. It took four days. He was cooperative when found, as most of the Massengill sales staff were. But not all. A Massengill salesman in Texas was so "thoroughly uncooperative," the FDA agent in charge of the search reported, that he had to ask the help of state police to persuade the salesman to divulge names.

Those were the exceptions. Most people were eager to help. Physicians especially were shocked to find that they might have prescribed poison to their patients, and most did everything they could to assist the FDA agents. One physician in the Appalachian region postponed his wedding to help search for a three-year-old boy whose family had moved into the mountain country after buying a bottle of elixir.

Massengill was still telling anyone who would listen that his firm had broken no laws, but Watkins, his chief chemist, was no longer certain they were blameless. A few days after taking his own medicine, he fell ill. On October 20, he wired the American Medical Association in Chicago, asking if they knew of any antidote for diethylene glycol poisoning. In a terse reply, the AMA gave the only answer it could: There was none known.

On October 22, the *New Orleans State* newspaper published a letter from Dr. A. S. Calhoun in Covington County. "Nobody but Almighty God and I can know what I have been through in these past few days," he wrote. He had prescribed the elixir twelve times before finding out about its toxicity. Now the doctor was suffering "mental and spiritual agony as I did not believe a human being could undergo and survive. I have spent hours on my knees," he wrote. "I have known hours when death for me would be a welcome relief." Six of his patients had survived, although "why they are not dead," he wrote, "I do not understand. For some obscure physical reason, their bodies were able to throw off the poisonous effects of the medicine. It is miraculous to me. I do not understand it. But I am grateful to Almighty God." The other six, however, were dead. One of them had been his best friend.

The same day that letter ran in Louisiana, Massengill released a statement to the press, reading, in part, "My chemists and I deeply regret the fatal results, but there was no error in the manufacture of the product. We have been supplying legitimate professional demand and not once could have foreseen the unlooked-for results. I do not feel that there was any responsibility on our part. The chemical sulfanilamide had been approved for use and has been used in large quantitites in other forms, and now its many bad effects are developing."

By October 26, there were thirty-six confirmed deaths. A twenty-five-year-old man in Mississippi died following several days of agony after taking the elixir to treat gonorrhea. A six-year-old boy in San Francisco died after taking it for an infected throat. A small child died in Arkansas. The American Medical Association was receiving frantic inquiries at the rate of one every five minutes. "Nothing like the present situation," Morris Fishbein said, "has arisen in American medicine in many years."

It appeared that the crisis was reaching its peak. Despite all obstacles, FDA agents had succeeded in recovering or accounting for more than 99 percent of the 240 gallons of elixir originally produced. Confirmed deaths from the remainder were continuing to come in. There were young men in small towns who had gotten prescriptions, perhaps for venereal disease, then gone up in the hills hunting. There were children who were still hanging on after a week. There were a few bottles still left in medicine cabinets. Judging by the patterns of death, the poor and the illiterate were still most at danger. The ones farthest from newspapers and hospitals. The ones hardest to find.

The massive publicity had an unintended side effect: Patients who needed sulfa drugs were now afraid to take them. All of Domagk's years of work at Bayer, all the meticulous testing, all the proven powers of sulfa were in danger of being undone by one careless chemist. The FDA and AMA responded by stressing in every public statement that it was almost certain that the cause of the poisoning was diethylene glycol, not sulfa. Sulfa, they repeated, remained a "very valuable" drug for the treatment of disease.

At the end of October, there were sixty-seven confirmed deaths, with a score more suspected. All but a few fluid ounces of the elixir was now accounted for, either swallowed by patients or recovered by the FDA.

On November 3, Samuel E. Massengill wrote the American Medical Association, stating, "I have violated no law."

Eight months later Massengill chief chemist Harold Watkins put a gun to his head and killed himself.

CHAPTER EIGHTEEN

On November 25, 1937, with death reports still trickling in, secretary of agriculture and future vice president of the United States Henry Wallace was called to Capitol Hill. America's legislators had requested a detailed report on the elixir tragedy, and Wallace himself, head of the department that included the FDA, intended to do more than deliver it. He planned to ask for action. A total of 353 patients had taken the elixir, he told the congressmen. "At least" 73 were dead as a result, and 20 more suspected fatalities remained to be investigated (the death toll would eventually top 100). Many more would have died, he concluded, if not for the quick response of health officials and the nausea-inducing effects of the elixir itself. If all 240 gallons of Massengill's tainted medicine had been consumed as directed, his people calculated, more than 4,500 people would have died. The FDA's speedy work was singled out for special praise. The report included details on the difficulties of the search, recounted the horrors of a number of individual cases, and included the complete texts of two letters: one from the New Orleans–area doctor who had killed his best friend and another written by Mrs. Maise Nidiffer of Tulsa, Oklahoma, the distraught mother of a young girl who had died. The secretary of agriculture took particular care in describing the actions and practices of the Massengill Company, including detailed information on each of the three cases in which the company or Watkins had been found guilty of violating the law. A copy of the elixir's label was appended with the note, "It will be observed that the preparation is a semisecret one, that the presence of diethylene glycol is not disclosed, and that no warning of danger appears."

The statistics and bureaucratic language could not hide a strong sense of outrage. In all, Wallace's report was, the *New York Times* noted, "unusually outspoken for a federal document."

The facts of the tragedy were a preamble to Wallace's main purpose: pushing for a complete overhaul of the nation's drug laws, specifically the obsolete 1906 Food and Drug Act. He concluded by offering Congress a blueprint for reform, ways to avoid such catastrophes in the future. His language was tough and to the point. He recommended changes that patent-drug makers had fought for years: requirements that all drugs undergo "experimental and clinical tests" for safety; that all drugs shown to be dangerous to health be removed from the market; and that labels on medicines list every ingredient, warn against side effects, and offer directions for proper use.

His action was directed at fraudulent patent medicines, but it was spurred by sulfa, the world's safest, most powerful, and effective curative, wrongly mixed with an industrial solvent. Now sulfa would do its part in breaking the back of the patent-medicine industry.

Wallace would not get everything he wanted, but his report would help push forward the greatest change in drug laws in history. His recommendations echoed those of Walter Campbell at the FDA. But these seemingly reasonable ideas had been blocked in Congress for decades. On one side were the makers of medicines, mostly patent-medicine makers, their associations, and lobbyists; on the other was the FDA, a loose coalition of consumer and women's groups, physicians, a few crusading journalists, and New Deal progressives interested in expanding federal oversight of a woefully underregulated field.

DRUG REGULATIONS in the United States were based on the Pure Food and Drug Act of 1906, a Progressive Era package designed to clean up conditions like those uncovered in Upton Sinclair's horror story about meatpacking, *The Jungle*. The law was weak, outdated, and aimed more at food than drugs. Even with minor alterations over the years, the law in 1937 was little changed. It did not require that any new drug be tested for safety on animals or in humans before being offered for sale. It did not require that a drug be proved effective. Almost any drug, as long as it was not a narcotic, could be sold without a prescription. There was no requirement that labels list all ingredients, proper dosages, or side effects. The only regulation of any real importance for drug safety dealt with advertising: According to the 1906 law, it was illegal to make false claims on a package. Even that applied only to the label on the bottle or the box around it. Drugmakers were free to say anything they wanted on flyers or in newspaper or radio advertisements. This small bite in the otherwise toothless package was aimed specifically at the makers of patent medicines, whose claims had become so incredible that it was felt necessary to invoke federal law to hold back a flood of lies.

Patent medicines in the early part of the twentieth century were as firmly established a part of American culture as jazz or baseball. Americans were accustomed to medicating themselves, deciding on their own treatments, and buying their own drugs. It went against the grain to have some doctor or federal agency telling Americans how to cure themselves. They were in the habit of going to their local druggists—independent businessmen of variable ability who compounded many of their own medicines—talking over their complaint, and getting what they needed, when they needed it, at a price they were willing to pay, without a prescription.

The ingredients, as well as many multi-ingredient medicines, were available either from "ethical" pharmaceutical and chemical companies—the elite of the pharmaceutical world, upper-end firms with high standards of quality that sold their compounds directly to physicians and pharmacies, disclosed ingredients and patents, and generally sold without direct advertising to the public—or from "proprietary" drugmakers, which did none of these things. The more common name for proprietary drugs was patent medicines. Manufacturers in this field jealously guarded secret recipes and sold their products directly to the public through massive advertising campaigns. They were masters of ballyhoo, filling every newspaper and papering every town with claims for the most amazing cures attributed to concoctions often brewed from the most worthless ingredients.

Patent medicines were a huge business. In the early 1930s, one dollar in ten spent by patients on U.S. health care was spent on patent medicines—enough to buy three or four bottles of unregulated drugs a year for every man, woman, and child. Estimates vary, but one industry insider reckoned that pharmaceuticals were the fourth-largest industry in the nation by the mid-1930s, with patent medicines accounting for half the sales. More than a billion dollars went into patent medicines every year, even during the worst of the Depression.

The makers of these medicines formed an entrenched, successful, wealthy set of interests with no hesitation about disarming any attempts to infringe on their business. For decades after 1906, lobbyists for the patent-drug makers effectively quashed any attempts to beef up U.S. drug laws and did their best to roll back the few laws that existed. In 1912, for instance, an amendment to the Pure Food and Drug Act was passed that limited its one bit of strength in relation to drugs: the provision that prohibited false and fraudulent advertising on packaging. The new amendment put the burden of proof on the government. If a drugmaker claimed that its magical water could cure cancer, rather than requiring that the maker prove it could, the new amendment forced the government to prove that it could not. This was, as one historian noted, "an exceedingly difficult task." In essence the government had to prove that the firm had knowingly made false statements with the intention of defrauding the public. And the patent-medicine makers could still say whatever they wanted in other advertising. It was open season on consumers. And many patent-medicine makers took full advantage.

Americans bought Lydia Pinkham's Vegetable Compound for female disorders, Marmola for obesity, Dr. Morse's Indian Root Pills for boils and worms. Brane Fude, Crazy Crystals, and innumerable other patent medicines for everything from diabetes to dementia. B&M External Remedy, a horse liniment, was sold as a tuberculosis cure. Radithor, a potentially lethal radium-laced water, was sold at a dollar a bottle to treat everything from persistent pain to cancer. The laws were weak, the profits were huge, and the stage was set for disaster well before the first bottle of Massengill's elixir was sold.

A few days after Franklin Roosevelt's first inauguration in 1933, FDA head Campbell buttonholed Rexford Tugwell, a handsome young incoming member of FDR's "brain trust" and assistant secretary of agriculture, and gave him an earful about the need for tougher drug laws. By that afternoon Tugwell was assuring him that the president was in favor of updating the 1906 act. The New Dealers made it part of the legislative agenda for their first one hundred days. The resulting bill, a sweeping reform that included almost everything Walter Campbell wanted, was introduced to Congress a few weeks later and quickly turned into a legislative disaster. It was too much, too soon. Tugwell, a Columbia University economics professor, was considered one of the more radical New Dealers, a proponent of a planned economy, an admirer of at least some parts of the Russian experiment—a leftist so unabashed that his political opponents were soon comparing him to Stalin. The patent-medicine makers immediately dubbed the reform legislation the "Tugwell Bill" in order to sink it. The bill's opponents included not only the manufacturers but their lobbyists, advertising firms, wholesalers, retailers, legislators representing southern and border states where patent-medicine firms were often headquartered—and many voters who began to identify attempts to reform the drug law with what they considered infringements on their basic right to self-medicate. The patent-medicine makers became the first organized business group to openly declare war on the Roosevelt administration. They began eviscerating the Tugwell Bill.

Campbell and his allies in the fight for drug-law reform—consumer groups, women's groups, some physicians—did their best to fight back. Tugwell, now a political liability, politely retired to the background and soon returned to academia. The fight for stronger drug laws would go forward under a new standard-bearer: New York senator Royal S. Copeland.

Copeland was an unlikely champion. Debonair, jaunty, a fresh red carnation always in his lapel, a physician with a flair for politics, he was something of a joke in the Senate, best known for trying to get his fellow legislators to exercise and eat better and for leading an effort to install "manufactured weather"—a form of primitive air-conditioning—in the legislative chambers on Capitol Hill. Public speaking was not his forte. Whenever he rose to deliver a speech in Congress, so many fellow senators left the chamber that a newspaper wag dubbed him "the greatest anesthetist of the Senate." The Senate pages nicknamed him "General Exodus." He did not even get along with his own political party. Copeland was a conservative Democrat who had no qualms about voting against Roosevelt's pet bills (FDR in return refused to endorse Copeland at reelection time). Copeland was better known for compromising with Republicans and business lobbyists than he was for fighting against them.

He hardly seemed the man to lead a controversial and much-opposed reform effort. Copeland, however, had his serious side as well. For one thing, he knew medicine. He was a homeopathic physician, trained in a form of what would now be called alternative medicine. In the 1930s homeopathy was a dying art, practiced by fewer physicians every year and denigrated by the American Medical Association as little more than quackery. Copeland was undaunted. He wrote a health column for the Hearst newspapers, hosted his own radio show, and promoted the healthful effects of a diet rich in bread and whole milk. He also made substantial sums of money by lending his name to medicine makers, including the promotion of patent medicines like Pluto Water. Underneath his lightweight façade was a deep commitment to public health. He had been health commissioner of New York City when an outbreak of botulism killed a number of citizens; the FDA had traced the source of the tainted food back to an olive-packing plant, and Copeland had learned to respect both the agency and the man who ran it, Walter Campbell. More important, Copeland was an able and patient politician, with an ability to pull conflicting sides together and get them to talk. He knew business and he knew medicine, understood the needs of the patent-medicine industry, and believed in building a stronger FDA to rein in its worst abuses.

The hearings he ran for the Tugwell Bill, however, turned into a fiasco. The patent-medicine lobbyists pulled out all the stops, kept the debate centered on Tugwell as long as they could, belittled what they called the "professor's bill," warned that drugstores would be "sovietized," alerted the media that drug advertising would be "wiped out in five years," and predicted that if passed the new law would turn the FDA into "a powerful, sinister machine" wielding the "heavy, cold, clammy hand of bureaucracy" over hardworking American businesses. "I have never in my life," testified a patent-medicine lawyer, "read a bill or heard of a bill so grotesque in its terms, evil in its purpose, and vicious in its possible consequences."

The Tugwell Bill was blackened, bloodied, and locked in committee. But in the process Senator Copeland got some backbone. He would not let go of what looked like a lost cause. He refused to let the bill die, reworking it again and again, calling meeting after meeting, making compromise after compromise. He would tell the pro-reform forces one day that "every slimy serpent of a vile manufacturer of patent medicine is right now working his wriggling way around the Capitol" trying to kill the bill. The next day he would listen sympathetically to a group of patent-medicine lobbyists. Perhaps because of his conservatism, he was able to keep businessmen at the table. Perhaps because he himself had used advertising, he was able to talk with lobbyists and advertisers. The anti-reform forces began working with Copeland as much as against him. Why not? They could talk until doomsday. Roosevelt had moved on to bigger issues; there was scant chance any major drug-reform bill would pass no matter how much they conferred. By 1935 the cause was so lost that patent-medicine lobbyists were able to bull through an amendment that would completely strip the FDA of what little power it had to regulate pharmaceutical advertising, giving oversight instead to the Federal Trade Commission. The result would have been worse than the 1906 law. Copeland's drug bill was now, as one despairing FDA staffer put it, "battered almost beyond recognition."

Copeland had some help, notably Campbell and his FDA. While federal agencies were not supposed to lobby for bills, Campbell figured that did not prevent the FDA from mounting "educational" exhibits. Instead of a teacher, he hired a savvy young advertising executive, Ruth Lamb, to head the agency's new educational office. Lamb realized that as long as the fight for drug reform was played out in back rooms, the powerful patent-medicine lobbies would win. So she moved the fight into the public arena. She made sure that Campbell spoke widely on radio and often to reporters. She made sure he repeated some simple points again and again: Unregulated drugs were "a vast field of pure cheat," and a new drug law would not limit free enterprise but rather ensure the safety of Americans. Lamb funneled plenty of pro-reform information to media, civic, medical, and consumer groups. Her office's greatest achievement, however, was the creation of a traveling exhibit that newspapers referred to as the "Chamber of Horrors." It started as a set of posters for making points at congressional hearings, each focusing on a different patent-medicine disaster. One showed a horrible death in Pittsburgh caused by Radithor, the radioactive water. Another displayed before-and-after pictures of an attractive Ohio woman who had been blinded, her eyeballs corroded by a cosmetic called Lash Lure. The simple point was that the 1906 law was unable to prevent these tragedies.

Lamb made sure that her posters were widely seen. A set was sent to Chicago for display at the Century of Progress Exposition. Others went on tour, displayed by pro-reform groups around the nation. Eleanor Roosevelt showed them to a group of congressional wives and talked about the Chamber of Horrors in her news column. Copeland brought the posters along every time he trekked to Capitol Hill. Then Lamb, working on her own time, expanded the exhibit into a book, *The American Chamber of Horrors*, published in 1936.

All these efforts failed to raise either a great outcry from the public or a grassroots movement to push forward new legislation. Most people were too concerned about simply making it through the Depression to worry much about a few idiots drinking themselves to death with radium water. Nor did the media hit the issue with high-profile muckraking cover stories. In fact, with few exceptions, the media stayed clear of the issue. Patent-medicine advertising was too important a source of income for newspapers and magazines. Not even FDR, regardless of his wife's interest, would speak strongly in favor of changes that might alienate his political base in the South. Despite what one observer later called Copeland's "enormous and intricate politicking," by the end of 1937 the latest version of his much-watered-down food- and drug-reform bill was still bottled up in committee. This time it looked like it might stay there for good.

The elixir tragedy changed all that. News coverage of the mounting death toll and the efforts of the FDA to contain the poisonings did more to change public opinion in several weeks than all of Ruth Lamb's educational efforts had done in several years. Congress was in recess when news reports about dying children and grieving mothers started headlining in October, but Walter Campbell was on the job, quickly using the tragedy to hammer home his ideas about drug reform. In his first public statement about the deaths in Tulsa, the FDA head linked the disaster directly to Copeland's languishing bill. As the death count grew, he kept up the pressure for change.

At last change would come. By the time Secretary of Agriculture Henry Wallace made his report to Congress on the elixir tragedy, politicians in every district were reading newspaper editorials, telegrams, and letters from constituents, all urging stronger drug laws to prevent more deaths. The greatest number of elixir deaths were concentrated in southern and border states, the same states that were home to many of the most powerful patent-medicine manufacturers and home states for some of the most vocal political defenders of the patent-medicine industry. Now, with public outcry particularly shrill in those districts, even formerly hard-line anti-reform legislators began softening their opposition. A number switched sides. Finally a few drugmakers—at least some of the most reputable among them—began to get aboard the pro-reform bandwagon. A week after the Wallace report, a vice president of Bristol-Myers, an ethical drug manufacturer, stood before the annual convention of the National Association of Insecticide and Disinfectant Manufacturers at the Biltmore Hotel in New York City and urged government action to prevent "such calamities" as had been caused by Massengill's "concoction." The American Medical Association joined the chorus, the editor of the group's *Journal* condemning Massengill's elixir as "apparently hastily rushed to market to meet an over-enthusiastic reception of a new remedy. . . . This tragic experience should be a final warning to physicians relative to the prescribing and administration of semi-secret, unstandardized preparations."

Groups ranging from the League of Women Voters to the American Association of Public Health, energized by the elixir news, began demanding not just that Copeland's long-delayed reform bill be passed but that it be greatly strengthened. California and New York moved to make sulfa available only by prescription. President Roosevelt called for a special session of Congress to consider proposed food and drug legislation. The result was SB 3073, the latest—and much more powerful—version of Copeland's drug-reform bill. The new bill put back many elements that had been stripped out during the previous few years of negotiations. Manufacturers were again required to submit information on safety testing, ingredients, and manufacturing processes and to provide any labels and samples that the secretary of agriculture might request. Approval to sell a new drug would be granted by the FDA. The newly strengthened bill then went further than anyone had dreamed possible; Copeland and Campbell got almost everything they wanted. One major exception was power over advertising, which again, in the new bill, was given to the Federal Trade Commission rather than the FDA—prompting Copeland to comment that "the consumers of this country are being raped"—but lobbyists and anti-reform politicians were suddenly ready to let that ride in order to pass the larger package.

On June 2, 1938, less than eight months after the first elixir deaths hit Tulsa, Congress passed a new Federal Food, Drug and Cosmetic Act. It was a milestone in legislative history. For the first time in U.S. history, it was required that new drugs be proved safe *before* they reached the market, that all active ingredients be listed on labels, and that warnings about misuse be provided to the users. The act stiffened penalties for drug companies, made it easier to stop false claims in advertising, and made it more difficult for people to buy whatever they wanted over the counter. It included, for the first time, cosmetics as well as drugs. Sales and production could be stopped if a medicine was found to be injurious to health. The new law created the modern FDA and, much amended, is still the cornerstone of today's pharmaceutical practices.

"We now have a law of which we can be proud," Copeland said. "It marks a very great advance, probably beyond that of any other country in the world." He was right. The 1938 act became a model for every other developed nation. Today's far more complex and comprehensive drug laws worldwide can trace their lineage directly to Royal S. Copeland and Walter G. Campbell. Less directly, the overhaul had started with the illness of the president's son, which kicked off a sulfa sales craze, which drew manufacturers eager to get in on the action, which led to quick formulations like the elixir, which resulted in a mass poisoning, which mobilized the public to back efforts to strengthen the laws—efforts that finally, at the end of a great circle, resulted in the passage of a legal reform that the president had wanted years before his son ever fell ill.

It was a great personal victory for Copeland, but it came at a price. Two weeks after Congress sent the new law to the White House for FDR's signature, Copeland found himself one morning unable to get out of bed. He told his wife that he just could not open his eyes. That afternoon he slipped into a coma. By evening he was dead. The cause was officially listed as circulatory collapse. His physician said that Copeland "had been a victim of the overwork and congressional strain against which he had cautioned his colleagues." In order to give the world a new model for drug regulation, the dapper homeopath, the man many had considered a joke, had worked himself to death.

A week later the Copeland bill was signed into law by President Roosevelt. The last major piece of New Deal legislation ever passed, it permanently and profoundly changed the relationship between government and the pharmaceutical industry and created the systems for drug development, testing, and oversight we use today.

Its impact went far beyond consumer safety. It reflected, regulated, and helped spur fundamental changes in how new drugs were discovered and developed. The pessimism that had pervaded drug research in the 1920s was changed, thanks to sulfa, to a belief that any research operation large enough, staffed with chemists talented enough, and directed rationally toward the cure of specific diseases could come up with breakthrough medicines and potentially huge profits. As a result all the major drug companies in the United States either established or dramatically increased the size of their research labs in the 1930s. Soon they were rolling out a dizzying series of powerful new medications.

These medications in turn had to be regulated for the public safety. The 1938 act firmly vested the responsibility for policing and enforcement of the law primarily with the Food and Drug Administration, which changed from a minor policing agency involved mostly with chasing down tainted food into a full-fledged regulatory agency charged with evaluating the safety and efficacy of all new drugs. The drugmakers who were to thrive in this new regulatory environment were those that matched the FDA's expectations by running their businesses like scientific enterprises. Hörlein's vision of pharmaceutical research ramped up to industrial scale became the model for the industry.

That translated into the decline of the once-powerful patent-medicine makers. It happened not only because the FDA was now able to shut down patent-drug makers for selling unsafe or ineffective drugs (although some of that did occur). It happened because the patent-medicine makers were dinosaurs in a new age of very expensive, long-term, scientific drug development. They simply could not keep up. The old patent-medicine firms stopped claiming to cure cancer and started focusing on over-the-counter medicines, cold remedies, and supplements. Those that could not adapt went under. It was the end of Dr. Gilbert's Sure Pile Cure, Gloria Tonic Tablets, Persian Pills, and Swift's Specific. It was the beginning of the era of antibiotics.

EVEN WITH THE new law in place, Samuel Massengill continued to maintain that he had violated no law. James B. Frazier, U.S. district attorney for Chattanooga, Tennessee, disagreed. He was not about to let any business in his state kill more than a hundred people without incurring some form of penalty. In June 1938, acting on the recommendation of the Department of Agriculture, he began criminal proceedings against the Massengill Company. The company could not be cited on the basis of any of the new laws; it could only be charged under the statutes in effect when the poisoning had taken place. So Frazier charged the drug maker with violations of the 1906 Pure Food and Drug Act, the most serious of which involved mislabeling and misbranding. He added a twist of his own, that "the elixir was sold as being fit for human consumption when, as a matter of fact, it was unfit."

Samuel Evans Massengill, although unrepentant, was eventually persuaded by his lawyers that it would be better to plead guilty. This he did. No one went to jail, but the company paid the largest penalty ever levied for violation of the 1906 act: $26,000—less than $250 per death. Massengill then released a brochure, *The Facts About Elixir Sulfanilamide.* In it he denied any errors in compounding the medicine and noted that there was still much doubt about whether diethylene glycol was actually the toxic agent involved. He spent the next several years gathering material and writing his *Sketch of Medicine and Pharmacy and a View of Its Progress by the Massengill Family from the Fifteenth to the Twentieth Centuries.* The book, which celebrated his family's dedication to medicine, did not mention the elixir incident.

CHAPTER NINETEEN

In early 1939, in a small, crude hospital far south of Khartoum in the Anglo-Egyptian Sudan, three British physicians waited for the inevitable. Every year for the previous four, an epidemic of meningitis had swept through tribes in the region, killing two of every three people it attacked and taking the lives of fifteen thousand Sudanese. The 1939 epidemic, the fifth in a row, "started out more violently than any of its predecessors," one of the physicians wrote, resulting in the deaths of thirty-three of its first forty-one victims. From their base health clinic in Wau, the area's largest town, the physicians trekked through the countryside, treating any victims they could find alive, counseling others to isolate the patients and avoid infection. There was little else they could do. Meningitis was a vicious disease. The death rate had always been high, and nothing they did had much effect. The British physicians concentrated on nursing the sick and trying to limit the spread of the disease. The only thing different this year came in the form of three small sample bottles of sulfa that had been sent to their clinic for the treatment of strep diseases and pneumonia. Two of the bottles had smashed in transit. But the third was intact. Strep diseases were not the problem of the moment in Wau. This meningitis was caused not by strep but by the more common cause, a related germ called meningococcus. Still, they had the new medicine, they had nothing else, and they had nothing to lose. Someone decided to try it on a meningitis patient.

There were problems from the start. The sick natives were feverish, often delirious. The medicine could not be given in tablet form, they wrote, "for it is not possible to administer drugs by mouth to a powerful African deprived of his reason by meningitis." So they crushed the tablets into powder, dissolved as much as they could, and shot it into a test group of meningitis patients. They had no way of knowing about proper dosage, so they guessed. There were twenty-one patients in the first group. The doctors hoped to save at least a few of them.

A few days later, all but one were still alive. The physicians immediately wired for more sulfa. Once it arrived, one of the British doctors stayed at the hospital while the other two went village to village, administering sulfa to every meningitis patient they could find. They asked the help of local "medicine men," as they called them, tribal healers whose dispensation was needed before the natives would accept treatment. The Sudanese healers knew how deadly the disease was. They told their people that the physicians had "magic in a bottle." They told them to take the shots. The physicians traveled day and night, injecting patients in grass huts, under trees, and along roadsides. The results, they wrote, were "spectacular." Within a few weeks, they treated more than four hundred patients. They saved more than 90 percent of them. They knocked out the epidemic before it could get started.

This was the largest human test of sulfa to date, and it was an unqualified success. The drug not only worked but worked against an unexpected disease. In previous tests pure sulfa had not been effective against meningococcal meningitis. But the sulfa the British tried was a new form recently developed by a British firm, May & Baker. Called M&B 693, it was intended primarily for use against pneumonia. The meningitis results were a bonus.

May & Baker was another of the European firms that had jumped into the sulfa game in 1936 as news of the German findings, Colebrook's success, and the French discovery of pure sulfa began to be known. Fourneau's and Domagk's labs were not the only ones tinkering with the sulfa molecule to improve it. May & Baker chemists, too—like many others in many other firms around the world—gave themselves the task of finding modified, patentable versions of sulfa that might be more effective against more kinds of bacteria, or might be less toxic, or preferably both. Azo dyes no longer played any role of importance. The sulfa molecule was attached to a variety of other side chains in hopes of finding a combination that would result in improved action. One observer likened the process to playing with a kite, fiddling with the tail, adding material or trimming it back to get it to fly better.

The French drugmaker Rhône-Poulenc, the firm with close ties to Fourneau's laboratory in Paris, owned a controlling interest in May & Baker. Soon after the Pasteur Institute's discovery of the power of pure sulfa, the British firm's small cadre of chemists was set to work searching for sulfa variations. When the British started their research in the spring of 1936, they, like everyone else, were working in the dark. Without knowledge of how sulfa stopped bacteria, there was no rational, plannable way to make the compound better. So they, like everyone, followed the German model, making educated guesses, using chemical intuition, and testing variant after variant in animals. Nothing interesting resulted for a year and a half. Then, in October 1937, a laboratory assistant got lucky. "Well, you know what would happen very much in those days," one of the May & Baker men remembered, "you would set out to try and make something, and you would find something on the shelf, and you would try it." This particular assistant found on the shelf a dusty bottle containing aminopyridine, a molecule prepared seven years earlier, for a chemist who had since left the firm, for a purpose no one could remember. The assistant worked aminopyridine into a form that could be attached properly to sulfanilamide, the correct bond was forged, and yet another sulfa derivative—this one numbered 693 in the May & Baker series—was forwarded for animal tests. Then the firm got lucky again. May & Baker's animal-test leader was on vacation. His facility was temporarily out of streptococcus-infected mice—sulfa testing seems to have created a temporary worldwide shortage of appropriate lab animals—so an assistant tried 693 against a related germ, pneumococcus. This was a particularly important bacterium, a top killer, cause of most of the world's cases of pneumonia. Pneumonia had been around as long as humans had, culling out the weak and elderly, "the captain of the men of death," one physician had called it. Pneumonia hit hard every winter. In December and January in the northern latitudes, hospitals would begin to see a rapid increase in patients appearing with chills and fever, weakness, a cough, and pain on one side of the chest. Physicians would listen with a stethoscope and hear wheezing or rattling in the lungs; a follow-up X-ray would show a shadow. After a week or two, the patients would develop a high fever, and then came the "crisis"—either a sudden improvement when the patients' immune response kicked in or, especially in older patients, a sudden decline, alarming exhaustion and debilitation, then death. Pneumococcal pneumonia (often called lobar pneumonia) killed hundreds of thousands of people every year. Pure sulfa worked well against the far rarer strep pneumonia but did not work well against lobar pneumonia. Nothing worked against it except serum therapy, which was far too expensive, difficult, and often dangerous for widespread use.

Then the pneumococcus animal test results with the British firm's new sulfa variation came back. The 693rd version had cured pneumonia in mice. The positive results continued as more mouse experiments were done, with sulfapyridine (the "official" name of the new drug, which never caught on in Britain, where everyone simply called it M&B 693) coming as a surprise. It worked on more than pneumonia. It also seemed to have a much stronger effect than plain sulfa against strep diseases, meningitis, staphylococcal, and gonococcal infections. Toxicity tests at the end of 1937 showed that it was almost as safe. By then some of the M&B research staff had tried it on themselves, too, without ill effect. Drug development in Great Britain was not dissimilar from that in any other country of the day, or the United States prior to the 1938 change in the laws. Testing was done relatively haphazardly compared to today's more stringently controlled clinical trials. By March 1938, M&B was being tested on patients. The first was a farmworker from Norfolk with severe lobar pneumonia. After treatment he fully recovered. The firm started large-scale production while more tests in pneumonia patients were still under way.

The company announced their animal test results ("striking success in the chemotherapy of pneumococcal infections") in *The Lancet* in May 1938; the positive results of the first large-scale clinical trial in humans was published five weeks later. The clinical investigators, a pair of physicians in Birmingham, found that the usual death rate for lobar pneumonia—27 percent in patients who received the best treatment of the day without sulfa—fell to 8 percent when they used M&B 693. Further comparative tests, they concluded, were unnecessary. Soon almost every physician in England was giving the drug to pneumonia patients. There was talk that the "ultimate cure" for pneumonia had been discovered. May & Baker put it on the market.

It took longer for it to catch on in the United States. The elixir tragedy was fresh in everyone's mind when word of this even more powerful sulfa drug hit. While May & Baker's new drug was put through animal testing at Johns Hopkins under the direction of Perrin Long and Eleanor Bliss, the British company made a deal with a U.S. firm, Merck, to market it in America. Merck in turn submitted it for approval to the newly empowered FDA in October 1938, noting a "great demand" for this fresh version of sulfa. But the FDA was in no hurry. The tougher approval regulations of the 1938 act were just beginning to be put into practice. Sulfapyridine was one of the first important drugs to be put through the revised system. The FDA wanted more toxicological data, more details on the metabolism of the drug, and more information on human tests before the agency would approve its use in the United States. The American Medical Association stepped in to help. Pneumonia season was about to start, which would translate into thousands of dead patients who might be saved with the new drug. A number of U.S. physicians had already secured the compound from European suppliers and started using it on patients, despite the lack of official approval. After hearing from Long and Bliss that their animal studies supported the British findings on the safety and effectiveness of sulfapyridine, the AMA moved to gather expert opinions from a hundred physicians who had already tried the new drug on nearly twenty-five hundred cases of pneumonia. They were careful to find everyone they could with no ties to Merck. Through the fall they collected, collated, and analyzed the information, as the number of deaths from pneumonia began to rise across the country. By December they concluded that the drug appeared safe, but they wanted more time to estimate proper dosages. No one was quite sure exactly what was required under the new drug law—how many new tests, how many patients, how safe was safe. No one wanted to make an error.

Meanwhile reports about sulfapyridine hit the press. *Collier's* magazine in December 1938 ran "Death to the Killer," a feature celebrating the end of pneumonia's reign, and the *New York Times* ran its own summary, "The Fight Against Pneumonia," the following January. The public began demanding the drug. A new pattern was set: the FDA carefully correlating test results and analyses, slowing the rush to approve a new drug, while patients and their families, excited about what they read and heard in the media—and, in this case, with the memory of the elixir tragedy fading—agitated for faster action. Some physicians, too, began to chafe under the stricter laws. Pneumonia cases were peaking, and sulfapyridine was still under review in February 1939 when a physician wrote Walter Campbell, head of the FDA, to say that he [the writer] was one of those doctors who had been all in favor of the 1938 drug-law overhaul. "I do not want to feel I was mistaken in doing this," he wrote, "but cannot help occasionally wondering if the present lack of a drug for seriously ill patients is not worse than operating under the old Food and Drug Act." Physicians were of two minds on the issue, many advocating faster action in the face of mounting deaths, others worried that the availability of this relatively unknown new drug would steer patients away from proven therapies—like serum—without adequate assurances of safety.

As spring approached, the AMA finally finished its survey work with a poll of forty-five physicians, a who's who of U.S. experts in infectious disease, almost all of whom recommended approving sulfapyridine. The AMA delivered its final verdict to the FDA: Overall, although there were reports of worrisome side effects, sulfapyridine was a "useful drug" and deserved approval. The 1938 act required that new-drug applications be approved or rejected by the agency within six months, a deadline fast approaching in March 1939 when the FDA finally gave its stamp of approval to the drug—as long as it was carefully labeled and administered "under close, continuous observation of a qualified practitioner of medicine." In other words, sulfapyridine would be available by prescription only.

The saga of sulfapyridine in the United States set a pattern for new-drug approval that would last until after World War II. It demonstrated to drug companies the sorts of tests that would be required before approval, the time it would take, and the earnestness of the FDA and AMA in making sure that nothing unsafe reached the market.

Sulfapyridine quickly made serum therapy for pneumonia a medical artifact. Shortly after its introduction, Lederle Laboratories, a top U.S. producer of pneumococcal serum, shut down its program and wrote off its investment in twenty-eight thousand serum-producing rabbits. Within a few years, sulfapyridine was saving the lives of more than thirty-three thousand pneumonia patients each year in the United States alone. The "captain of the men of death," for the first time in history, had been unseated.

If one sulfa variant could be found that worked this much better than pure sulfa, then others could as well. Drug firms kept up their research programs, looking for the next, more powerful, less toxic, patentable form of sulfa.

They found it in 1939, in the form of sulfathiazole, then again in 1940 in sulfadiazine, the first more powerful than sulfapyridine, the second less toxic. "Instead of feeling miserable, as with the other sulfa compounds, patients who have taken sulfadiazine feel fine next day and start asking for food," Perrin Long said. These were the best of a huge number of sulfa derivatives made during the late 1930s and early 1940s. By 1942, no fewer than thirty-six hundred sulfa derivatives had been synthesized and studied, and more than thirty of them were being sold in the United States under different names (producing what one observer called "a remarkable confusion in the literature"). But the big four variants—sulfapyridine, sulfathiazole, sulfadiazine, and later sulfaguanidine—would dominate the world market. The French, as Fourneau lamented, were falling behind in the game. Domagk's laboratory at Bayer, too, was working on more effective versions, but its global dominance of the sulfa market was over. Germany's Prontosil and France's Rubiazol as well as pure sulfa were all eclipsed. Each new sulfa variant upped the ante on effectiveness; each newer version tended to supplant the older. Sulfapyridine was, gram for gram, eighteen times more powerful than Prontosil; sulfathiazole was fifty-four times more effective; sulfadiazine was one hundred times more effective. It appeared that there was no end to the possibilities.

The number of diseases that could be treated with the new sulfa drugs was also growing. Sulfapyridine (M&B 693) extended sulfa drug treatment from strep to pneumonia and bettered results against gonorrhea and meningitis. Sulfathiazole worked just as well or better against all those diseases, and it added staphylococcal infections to the list. Although a major sales advantage for sulfadiazine was its remarkable nontoxicity, it was also more effective against strep than anything seen before—Domagk, after testing it, was said to have joked with his Bayer team, "I really ought to advise you to stop looking for further drugs. As I am able to cure a hundred percent of my infected mice with minimal doses, I cannot see anything better." Sulfadiazine looked like the king of sulfa drugs. But sulfaguanidine, which had the unique quality of being retained longer in the gut than any other, extended the power of sulfa to an unexpected disease, bacterial dysentery, another common killer, a huge health problem worldwide, and a malady that had not responded to any previous sulfa variant.

The future seemed limitless. The only thing missing was an explanation of how the medicine worked. Then finally that, too, was discovered, eight years after Klarer made the first molecule of Prontosil. Donald Woods and Paul Fildes, a pair of London researchers, found that sulfa was less a magic bullet than a clever impostor. They started from the observation that sulfa never worked as well when there was dead tissue around, or a lot of pus, as in uncleaned wounds. Why? Some researchers thought that the dead tissue might release some substance that interfered with sulfa's action. Woods went looking for the mysterious anti-sulfa substance and found it in yeast extracts. A small amount of it, whatever it was, had "astonishing" anti-sulfa abilities. Fildes helped characterize the mysterious substance. It was a small molecule, just as sulfa was a small molecule. In fact, the mystery molecule looked like sulfa's twin in terms of size and chemical character. They determined that it was para-aminobenzoic acid (PABA), a chemical involved in bacterial metabolism, a sort of foodstuff necessary for bacterial growth (it plays a role in human nutrition as well, although it is better known as a sunscreen ingredient). Some bacteria can make their own PABA. But others cannot and have to find it in their environment. For these bacteria PABA is an essential metabolite. If they cannot find it, they starve.

Woods and Fildes, through a series of solid, well-designed experiments, proved that sulfa worked because it looked a lot like PABA. When sulfa was around, the bacteria would try to metabolize it instead of PABA. The two molecules were so similar that an enzyme critical in keeping the bacteria healthy would mistake one for the other, binding sulfa instead of PABA. But the sulfa could not be metabolized. The enzyme, with sulfa stuck to it, became useless. The bacterial machinery, gummed up, would slow to a stop. The bacteria, denied a nutrient they needed, eventually starved to death. "What has been looked upon as differing degrees of vulnerability among bacteria" to sulfa, Woods and Fildes wrote, "may actually be differences in the quantity of PABA produced by the several species . . . or to difference in the importance of PABA in their metabolism." Sulfa worked brilliantly on cocci because they could not make enough of their own PABA; additional PABA released from dying tissue or found in pus hindered sulfa's action because it provided the sensitive bacteria the food they needed. This surprising finding opened up a whole new field of research devoted to finding more impostor molecules like sulfa that could substitute for needed bacterial foodstuffs. The "antimetabolite" approach that began with sulfa would yield a host of new therapies, including a number of drugs like methotrexate and 5-fluorouracil, used to treat cancer.

Now researchers knew how sulfa worked and why it worked on some bacteria (those that required PABA from their environment) and not others (those that made their own). By 1940 it was also becoming clearer which diseases sulfa drugs could stop and which they could not. One form of sulfa or another was now standard therapy against pneumonia, childbed fever, erysipelas, strep blood infections, and the most common kinds of meningitis. Sulfa stopped urinary tract infections, trachoma, chancroid, mastoiditis, otitis media, and a host of less common or less dangerous strep infections. Newer sulfas worked well against gonorrhea. Strep sore throat could be treated with some success; sulfa was especially good at preventing recurrences. Sulfa drugs were increasingly helpful against staphylococcal infections and dysentery. They did not work well, however, against bacterial diseases like typhoid fever, tuberculosis, anthrax, or cholera. Thanks to the success of sulfa, new and newly energized research on finding new medicines would seek cures for these and other remaining diseases. The results would not be long in coming.

The United States began enthusiastically adopting sulfa therapy not only for the treatable diseases listed above but for everything from appendicitis (the idea was to give the drug before an appendectomy; it did cut the incidence of appendicitis deaths in half) to the common cold (sulfa had little effect on any viral disease, but people took it anyway). During 1941, seventeen hundred tons of sulfa were produced in the United States alone, enough to treat millions of patients. The new drugs were saving more than fifty thousand lives in the United States each year. The numbers worldwide were double or triple that—production and use increasing at such a rapid rate that no one was able to keep careful track. Sulfa was, at the end of the 1930s, being sold in vast quantities. In America, even after the change in drug laws, you could still get sulfa from your corner pharmacist by the bottle or tablet by tablet. People used it whenever they felt ill, or they popped a couple of pills before going out for the evening to stave off gonorrhea. It was, of course, of no use to swallow a tablet or two for anything. Sulfa worked only if given in relatively large quantities over a period of time. Still, "A really great era of sulfa-miracles is dawning," enthused the *Reader's Digest* at the beginning of the 1940s. By the time that statement was made, however, a darker era had already begun.

CHAPTER TWENTY

In the late summer of 1939, Gerhard Domagk and his family vacationed in Dahmeshöved, a tiny fishing village on the Baltic coast. Domagk had loved the Baltic since his student days in Kiel. He and Gertrude had planned and saved for a way to get out of Elberfeld in the summers and finally invested in a small second house in this quiet hamlet. It was isolated and primitive, without running water or electricity, and it was beautiful. Domagk spent his time writing poetry and weeding in the garden, enjoying the clear air and sunsets, wondering exactly how many shades of blue there were in the sky and the sea, while his three children swam along the shore and roamed the nearby fields. This vacation, however, was less than idyllic. War was coming, and everyone in Germany knew it. Less than a year earlier, the Munich Agreement had given Czechoslovakia to Hitler. Now German troops had marched into Austria. There was talk of invading Poland. Everyone was worried about being called to military duty. Domagk, forty-three years old, was still young enough to be called back into medical service. His eldest son was almost old enough to become a soldier, his middle son too young yet, but if the war went on . . . Domagk's strong feelings of patriotism for his country were now tempered by his own experience of war. He compared Hitler's actions, and those of the other European leaders, to the behavior of wild beasts circling each other and growling even when there was plenty for all to eat. He did not care for the Nazis and their policies. But above all he was German, and he was prepared to do what his nation required.

"Officially, mobilization has not yet been announced," Domagk wrote in his journal, "but they are bringing in horses from the surrounding villages. One of the government buildings has a sign saying, 'We accept volunteers.' Compared with 1914, the mood is subdued. The memories of the dreadful war experiences are still awake in people's memories. The cafes and bars are full of young people, too young to go to war. They are playing dance music and hits, unable to hide a subdued determination. What is it all about? Is there no other solution?"

With tensions rising at the end of August, an explosive mine was found floating in the water near Dahmeshöved. Resorts along the coast emptied. Domagk decided to cut the vacation short, and his family returned to Elberfeld while he traveled to Berlin on business. He found the capital "unusually quiet," everyone deadly serious, very few private automobiles on the streets. He wanted badly to return home and, as soon as he could, arranged to leave the capital. Worried about crowds, he arrived two hours early at the train station, which he found packed with departing soldiers, Red Cross volunteers, red-eyed wives, and clinging children. Everyone, it seemed, was either under orders or trying to get out of the city before war was declared. "The trains are arriving," he wrote. "The panic starts. I help as much as I can with some children, screaming, whining, sighing. Dust, noise, children crying, marching boots, pushing, screaming, pulling, shouting, a whistle."

He made it home and began readying himself for service, buying a sturdy pair of boots, packing his military suitcase, comforting his wife and speaking with her about what might happen. A few days later, German troops invaded Poland. In turn, Poland, England, and France declared war on Germany. Blackout restrictions were put in place along the Rhine. Duisberg's giant flashing BAYER sign was switched off—it would never shine again—and the nation threw itself into what everyone again hoped would be a quick war.

Domagk was grief-stricken. The Nazi ideology held no attraction for him. The idea of crippling science because of racial and political views—the way Hitler had driven Jewish researchers out of Germany—was anathema. Domagk believed that the National Socialists had come to power by lying to Germans distraught about the conditions following the First World War, promising progress while delivering prejudice. The older he grew, the more he was attracted to pacifism. The new war threatened everything that he and other Germans had struggled to build. They had gone through terrible times—inflation, starvation, humiliation—and the nation, it seemed, was just now getting back to health. All that was in danger. He was especially concerned about the progress of medical research, lately so fruitful, now interrupted by mobilization and refocused research priorities. IG Farben was already deeply invested in making artificial rubber and gas for the army. Domagk's medical-research group at Bayer was asked to turn their attention from curing civilian disease to military needs, focusing on perfecting a mix of sulfa variations that could be sprinkled on battlefield wounds to fight gas gangrene.

After war was declared, Domagk fell into a mild depression. Then, in late October, he fell ill with the flu. He treated himself with Prontosil, which he still felt was among the best of the sulfa drugs when properly used. On the morning of October 26, he dragged himself into the laboratory and gave a tour to a group of visiting military doctors. In the afternoon he went home to rest. There he received an unusual phone call from a Swedish journalist in Berlin, asking detailed questions about his research and his awards to date. It was curious, given what was going on. Later that night the same reporter called back, this time with congratulations. Domagk sat stunned as the reporter told him that he had been awarded the 1939 Nobel Prize. The official notification, a telegram from the Professional Committee of the Royal Caroline Institute, arrived at midnight, followed by another phone call, this time from the Reich newspaper press in Berlin, asking for confirmation.

Domagk was thrilled. He was following in the footsteps of German medical heroes Koch and Ehrlich, both Nobel Prize winners in medicine or physiology. But his happiness was mitigated by anxiety. He knew, as every German scientist knew, that Hitler had forbidden any German citizen to accept a Nobel Prize. The decision had been made following the award of the 1935 Peace Prize to Carl von Ossietzky, a pacifist and critic of the Nazi regime. At the time of the award, Ossietzky was a prisoner in a German concentration camp. His award had followed a year of "suggestions" to the Nobel Committee from anti-Nazi activists throughout Europe. It was obvious that the Swedes were trying to make a political point. The German press was prohibited from mentioning Ossietzky's prize, and Hitler, stung by the criticism, decreed that henceforth no German would be allowed to receive a Nobel. Instead he created a substitute, an all-German, all-Nazi state prize for achievement. Ossietzky caught tuberculosis in the camps and died in 1938, but Hitler's ban was still in effect.

For three years after the Ossietzky affair, the Nobel committees (prizewinners are chosen by several committees in Sweden and Norway) did not award any prizes to Germans. Then in 1939 they gave three: a delayed award in chemistry to Richard Kuhn—an Austrian, but by then Austria had been annexed to Germany (Kuhn was officially the 1938 chemistry winner, but his prize was not announced until 1939); the current year's chemistry prize to Adolf Butenandt of Berlin; and the 1939 prize in medicine or physiology to Domagk.

Domagk's prize was awarded by a faculty committee at the Karolinska Institute, a leading Swedish medical school. Once it was clear that Domagk was in the running for the prize, the Karolinska committee's head, Folke Henschen, a pathologist, wrote personally to Reichsmarshall Hermann Göring asking for advice on the current status of the German Nobel ban—and perhaps hoping to stave off a potentially negative reaction from the Nazi government. He never received a reply. Henschen then asked the advice of the cultural attaché at the German embassy in Stockholm, a reasonable diplomat who believed that it was in the best interests of German science to lift the ban, a man who in turn wrote Berlin for instructions. The telegraphed reply read: PLEASE INFORM NOBEL COMMITTEE THAT THE AWARD OF NOBEL PRIZE TO A GERMAN IS POSITIVELY UNWELCOME. Henschen reported this to the Karolinska committee but continued to advocate for Domagk's award even without the support of the German government, arguing that the prestige of the Nobel Prizes for science would suffer if it appeared that the committee was being swayed by politics.

Domagk, he said, clearly deserved the prize on scientific grounds. The committee agreed unanimously. The next day the award was announced to the world.

Domagk was first of the three German prizewinners to hear the news. He told no one at work other than his director, who asked him to tell no one else. It was a wonderful secret. He was elated while he awaited word on whether he could accept, and he was not surprised the next day, or the day after, when he found nothing in any German newspaper about the honor. In the absence of any word from the government, he decided to write to the chancellor of the university at Münster, where he still held his appointment, asking for advice and requesting that the chancellor check with the German Ministry of Education. A quick telegram in reply said only, WAIT FOR FURTHER INFORMATION. A week passed. When a worker at Elberfeld heard from an aunt in Finland about the honor, Domagk asked her to keep quiet. Then the chancellor told him that he had spoken with the Reich's Ministry for Home Affairs and had tried to make the point that the scientific prizes, like Domagk's, awarded by the Swedes, were quite distinct from the Peace Prize, awarded by the Norwegians—different enough to warrant Domagk an exemption from the Ossietzky policy. "As soon as I get more information I shall let you hear of it," the chancellor wrote. He signed below the common closing of the day: "Heil Hitler."

By now Domagk was beginning to receive congratulations from colleagues in nations where news of the prize was not censored. Feeling that it would be impolite to delay some sort of response to the Karolinska any longer—and still hoping for government approval—on November 3, Domagk sent a message thanking the members of the committee for the recognition and adding, "To the best of my knowledge, as a German national, I am not allowed to accept the prize. I am still in the process of obtaining detailed information about the legal position. . . . I cannot tell you at this moment in time if I shall be able to come on 10 December to Stockholm. I will let you know as soon as I have further news."

A few days after sending it, he was ordered by the German Ministry of Foreign Affairs to give them a copy of his note for their review. Then there were more days of silence. When the other two German Nobelists, Adolf Butenandt and Richard Kuhn, learned of their prizes, they were immediately contacted by the Ministry of Foreign Affairs and forbidden to acknowledge them. Later in the month, the two were summoned to the Ministry of Education in Berlin for "instructions." There they sat across a table from an SS brigadier and a second man they did not know. On the table were three typed, unsigned letters, one each for Butenandt and Kuhn, the third, presumably, for Domagk. Each one was addressed to the appropriate Nobel Committee; each curtly refused the prize. All of them said roughly the same thing: Not only was it against their nation's law for them to accept, but the awards also represented an attempt to provoke them to disobey the Führer. When Kuhn and Butenandt tried to make factual corrections, they were told that every word had been approved by Hitler personally. Nothing could be altered. They were told to sign the documents. They were asked to mail them from their hometowns. They both signed, they said later, because they were afraid for their own safety, for their wives and children, and for their research institutions. Both men went on to hold prestigious positions in German science through the war.

Domagk was not there with the other two scientists because he was in prison. On the evening of November 17, 1939, the Gestapo arrived at his home, searched it, arrested him, and took him to the Wuppertal jail. There he was held for questioning. Days went by. When he asked his jailers why he had been arrested, the only answer he received was that he had "been too polite to the Swedes." He asked for pen and paper and kept a journal of his stay. One day he wrote, "It is easier to destroy thousands of human lives than to save a single one." Another day, he remembered, "a man arrived to clean out my cell and asked me what I was doing there. When I told him I was in jail because I had received the Nobel Prize, he tapped his head and declared, 'This one is mad.' " There seemed to be no point to the endless questioning. Although Domagk did not record it in his journal, it is likely that he was asked to sign a document similar to Kuhn's and Butenandt's. If he was asked, he refused. But the experience began to grind him down. He suffered from feelings of intense anxiety. He started suffering from chest pains.

Exhausted and unnerved, he was released a week after his arrest. "My attitude to life and its ideals had been shattered," he wrote. Still shaky, he went to see his friend Philipp Klee at the Wuppertal hospital for a checkup. Klee was himself fighting to keep his Jewish wife safe. He told Domagk there was no physical reason for his continuing chest pains. He was suffering from mental stress.

A few days later, on his way to Berlin to deliver a scientific talk, Domagk heard his name called over the loudspeakers in the station at Potsdam. He was told to report to an exit. There he was met by a Gestapo officer who escorted him to the local offices of the secret police. He was told that he would not deliver his lecture as planned. Instead he was given a letter to the Nobel Committee and asked to sign it. The letter, similar to those already signed by Butenandt and Kuhn, declined the prize.

Domagk signed.

THE REFUSALS OF the three German prizewinners, added to the start of war in Europe, led to the cancellation of the 1939 Nobel Prize ceremonies. Each winner was to have received, in addition to the gold Nobel medal and public acclaim, a substantial amount of money—140,000 Swedish crowns—a rich award worth several years' salary for most scientists. Domagk and the two other German winners that year would never see it. Their official refusals left their prize money unclaimed; after a short time, it reverted back to the main prize fund, in accordance with the rules of the Nobel Foundation.

Domagk's Nobel Prize had turned from a great honor to a great burden, and the negative effects were not over. The prize was awarded to Domagk alone, crediting him properly with the "discovery of the antibacterial effects of Prontosil"—rather than for the discovery of Prontosil itself—and ignoring the chemists who had created and patented the compound, Fritz Mietzsch and Josef Klarer. When news of Domagk's award percolated through Bayer, it triggered a resurgence of bad feelings. Klarer, especially, resented being relegated to a footnote. Domagk's publications were often cited. Klarer's work was invisible. Domagk received scores of letters thanking him for his great discovery. Klarer received none. Domagk won the Nobel Prize. Klarer was not mentioned. Mietzsch was mature enough to hide any disappointment he might have felt—in any case, his contributions were fewer—but "Klarer held that the effects could not have been discovered without someone having pioneered the isolation of the sulfanilamides and submitted them for testing," wrote one of his colleagues. "Klarer was very disappointed and embittered." He never recovered. After the Nobel Prize eluded him, the volatile and brilliant chemist who created Prontosil, the greatest medicine on earth, lost interest in further work with sulfa. Within a few years, Klarer ceased his research in the field entirely, turned to less productive areas of research within Bayer, and disappeared from the world stage.

The Nobel Prize mechanism, designed to credit a single great discovery by a single person, was not built to properly credit industrial scientific research. In places like Bayer, discoveries were made cooperatively by large teams, with individuals filling specific needs. Giving a single person within this structure a Nobel was like awarding a football championship to one player. Certainly the discovery of Prontosil deserved a Nobel Prize. Certainly Domagk deserved his name on the prize. But, arguably, so did Klarer and Mietzsch, who had made the molecules. So, perhaps, did Hörlein, who had created the research structure in which the discovery was made. A strong case could be made for sharing the prize with Fourneau's French team, who discovered the active portion of the molecule and whose work led directly to new and productive areas of research: the idea of bioactivation, the discovery of the molecule's method of action, and the synthesis of stronger derivatives.

Making improved versions of sulfa drugs was now the top priority for Domagk's research group, as it was for many drugmaking firms around the world. Once the Gestapo got what it wanted, Domagk was put to work again looking for better and safer versions. He was still a bit in shock and still suffered from chest pains—as he would for the rest of his life—but he threw himself back into his research, focusing on the wound-infection issue and tuberculosis. Just after Hitler invaded Poland, he and his team came up with a new sulfa variation, Mesudin, more effective than anything seen so far in treating gas gangrene. Domagk took a special interest in spurring its use among German army physicians, hoping to ease some of the suffering he had seen at the field hospital in the Ukraine a quarter century earlier.

Reinhard Heydrich was thin, blond, and noted for his extraordinarily long fingers. People also noticed his eyes, which reminded them of a wild predator's. Heydrich was in 1942 the brightest rising star in the Nazi Party, only in his mid-thirties but already number two to Himmler in the SS; head of the Nazi secret police; architect of the fake attack on a German radio station that had been used as an excuse for invading Poland; organizer of the Einsatzgruppen (action squads) that massacred Russian Jews; a central designer of the "Final Solution" for the rest of them; most recently elevated to deputy reich director of Bohemia and Moravia. He was a personal favorite of Hitler's. Some of his enemies called him "the Blond Beast," others "the Butcher of Prague."

Luckily for his enemies, Heydrich was also supremely arrogant. That year he took to commuting between his home in the Czech countryside and his headquarters in Prague in an open Mercedes touring car without an armed escort. It was his way of thumbing his nose at the impotence of the local Czech resistance. His overconfidence was something that could be used against him.

Three Czech agents trained and armed in England parachuted into the area in May 1942 with a single assignment: kill Reinhard Heydrich. They hid with local Resistance members while they planned an ambush along the route of his morning commute, choosing a place where the road took a sharp turn and his driver would be forced to slow the car. One assassin was posted on the approach, his job to signal the other two with a pocket mirror as Heydrich's car neared. The next of the assassins down the road was to spray the car with bullets and the third to throw grenades. On the morning of the attempt, everything seemed perfect. Then, for some reason, Heydrich was later than usual. The two agents down the road became worried about an increasing amount of civilian traffic along the highway. Then they saw the mirror flash. One agent ran into the road with his Sten gun, planning to shoot the driver as they came around the corner. He raised his gun and pulled the trigger. It jammed. Heydrich's driver saw the agent and accelerated around him. Heydrich and his driver could simply have continued on and escaped, but instead the deputy reich director drew his pistol and ordered his driver to stop. He intended to go back and kill the man in the road. Then the third Czech assassin ran toward the car and threw a specially designed grenade, hoping to land it on the seat next to Heydrich. He missed. The device blew up under the car near a rear tire, spraying the Czech agent with gravel and shrapnel and blinding him. Heydrich struggled out of the car and began firing his pistol. One assassin dove for cover; the other ran off with Heydrich's driver on foot behind him. Then Heydrich collapsed in the road. The Czech bomb had blown shrapnel, bits of upholstery, and wire from the car seat deep into his abdomen. Heydrich had not bothered to armor the underside of his touring car. Help arrived, and he was rushed, badly wounded, to a hospital in Prague.

German physicians immediately performed surgery. Himmler, the head of the SS, rushed his personal physician—Karl Gebhardt, skilled surgeon, SS Gruppenführer, head physician for the Berlin Olympics, and Waffen SS medical department chief—to Prague to supervise the postoperation recovery. Gebhardt assessed the patient, decided that more surgery was required, oversaw it, then followed it with blood transfusions and sulfa. It looked at first as if they had saved Heydrich's life. As he recovered, German soldiers cornered the three Czech agents in the basement of a Prague church, and there the assassins, after exhausting their ammunition in a fierce gun battle, committed suicide.

A few days later, Heydrich's temperature started soaring. His wounds had become infected, and the infection was spreading to his blood. Gebhardt and the other physicians increased the sulfa and treated the intense pain with morphine, but it was too late. Heydrich fell into a coma and died early on the morning of June 4.

On the day of Heydrich's death, German soldiers executed 152 Jews in reprisal. His body was brought to Berlin for a hero's funeral. Hitler himself spoke with emotion at the service, stopping afterward to talk with Heydrich's young sons, telling them that their father would be always to him the "man with an iron heart." A week after Heydrich's death, on the morning of June 10, 1942, German troops stormed into the small Czech mining village of Lidice, a place they believed had sheltered the three assassins. They shot all the men, then gathered up the women and children and sent them to concentration camps. Building by building, they destroyed the town with explosives, then leveled the ruins, plowed the ground, and seeded it with grain. The name of Lidice was struck from all German maps.

Still saddened by Heydrich's death, Hitler began hearing hints from his own personal physician that Karl Gebhardt, the physician in charge of Heydrich's recovery, had let his patient die because he had relied too heavily on surgery and not enough on sulfa treatment. Properly used, sulfa, it was said, could have saved the life of Hitler's favorite. Gebhardt denied it. To clear his name and decide the question, within weeks of Heydrich's death the Germans set up a new medical-research program under Gebhardt's direction.

The women of Lidice were sent to Ravensbrück, a women's work camp ninety kilometers north of Berlin. There they joined thousands of other women, mostly from Poland, mostly political prisoners accused of working for the Resistance or for being Gypsies, homosexuals, leftists, Jews, or anti-Nazis. Many of them had already received death sentences. Every woman at Ravensbrück was given the same uniform: a single striped dress and strips of rag for shoes. Underwear and socks were rare luxuries. They rose at 4:00 A.M., drank imitation coffee, and stood at attention in the freezing streets between the barracks to await work assignments. They dug ditches, repaired shoes, and spun thread. The luckiest worked in the kitchens; there at least they had enough to eat. Many died of starvation or exposure. Good numbers are difficult to find for Ravensbrück; the best estimates are that about two-thirds of the more than one hundred thousand women who went there died there. Geneviève de Gaulle, the niece of the French general, a Resistance member, and a Ravensbrück prisoner, wrote, "As we entered the camp it was as though God had remained on the outside."

On July 27, 1942, less than two months after Heydrich's death, fifteen Ravensbrück prisoners, all of them Polish women, were taken to the commandant's headquarters. Their identification was verified, and their legs were examined by Herta Oberheuser, the camp's resident physician, a sharp-featured young woman who had taken the job—her first after medical training—because it paid better than most positions available to female doctors. She was assigned to assist Gebhardt, and she did as she was told. She was an ardent Nazi. After the examination the prisoners were escorted to the camp's medical clinic, where they were anesthetized and operations performed on their legs. This first group was treated carefully. Their legs were surgically opened, the blood vessels leading to the incisions were tied off, pure cultures of bacteria were smeared into the wounds, and they were surgically closed. The patients then received varying amounts of sulfa. The idea was to mimic the conditions that had killed Reinhard Heydrich.

Gebhardt oversaw the experiments, but the head of SS medical services, Ernst Grawitz, also took a personal interest, visiting the camp to check the progress of the study. Grawitz decided, after observing that none of the first group had died, that the procedure did not adequately replicate Heydrich's case. The tests were to be continued in a more "realistic" fashion. Further groups were brought in and operated upon. Now the German physicians forced dirt, slivers of glass, and wood shavings into their wounds, along with the appropriate bacteria. Bones were broken. Tissue was removed. Grafts were attempted. One report referred to the use of actual gunshots. After infection set in, surgery was performed and sulfa administered, following as closely as possible the series of events experienced in the Heydrich case. Always, to maintain adequate scientific control, a few subjects in each test group were not given sulfa. Soon women began dying.

Those who survived could be seen hopping and limping around

camp on makeshift crutches. The other inmates called them *Kaninchen,* German for rabbits. It was ironic: The same word appeared often in Domagk's notebooks (and those of everyone else who performed animal tests on drugs) abbreviated as *Kan*—test animals sacrificed to perfect new drugs.

As more and more Ravensbrück prisoners began to realize what was happening, the women started fighting back. When ordered to report to the clinic, some refused to go. Those who resisted were taken to "the Bunker," the most feared place in Ravensbrück, a prison-within-a-prison, a small complex of isolation cells and workrooms where Gebhardt and his assistants continued their research. Now, according to de Gaulle, the researchers were working without anesthesia or antiseptics. The number of *Kaninchen* grew. The other inmates pitied them and spared them what they could—scraps of food, blankets, underwear. They drew up a formal petition of protest and took it to the camp commandant, a small troop of women on crutches accompanied by those who supported them. The petition asked the commandant if he knew about the experiments, asked whether they were operated on as a result of sentences passed upon them, asked whether he knew that international law forbade this activity. They were not allowed to see the commandant and received no answer to their petition. In August 1943, when the Germans ordered another increase in the number of patients, the Polish barracks block went on strike. In spite of punishment, they refused to turn over any more women. The research stalled. As Allied forces neared the camp, the Nazis tried to round up and kill as many *Kaninchen* as they could find, but the women in the Polish block hid as many as they could. Fifty or so *Kaninchen* made it through the war. One of them, Jadwiga Dzido, displayed her scarred legs before judges at the Nuremberg Trials. Another, Vladislava Karolewska, testified about the nature of the experiments.

The sulfa experiments at Ravensbrück came fully to light at Nuremberg, one of a number of incidents in which Nazi physicians had used camp inmates in medical experiments. After the trials, Herta Oberheuser was sentenced to twenty years in prison (she was released after serving five years and worked in private practice until the late

1950s, when she was rediscovered and her license to practice medicine was revoked. She spent the rest of her professional life as a kitchen worker). Karl Gebhardt was hanged. Ernst Grawitz, the highest-ranking medical officer in the incident, the man who ordered increasingly severe tests, was never tried. He committed suicide in Berlin at the end of the war.

CHAPTER TWENTY-ONE

THE ASSASSINATION of Reinhard Heydrich also had an effect on Gerhard Domagk. In May 1942, as the Blond Beast lay dying in a Prague hospital, Domagk was still working to get the German army to embrace new sulfa variations that he believed would solve the problem of gas gangrene. Marfanil and Mesudin, as Bayer named the newest sulfa drugs, had been shown in Domagk's laboratory to be vastly more effective in stopping severe wound infections than anything else available. Gas gangrene was the affliction that had started Domagk down the road of antibacterials, the disease he had sworn to fight as a young medical assistant in the Ukraine. "The real birth of chemotherapy as far as I am concerned took place in the Great War of 1914–1918, when I swore an allegiance with my fallen comrades. Those were my first principles and they are still valid," he wrote in his journal during World War II. "They stand over my work like a shining star." When in 1939 the Bayer team first came up with sulfas that could stop gas gangrene—one of them the last sulfa variation that Klarer would ever provide—Domagk felt certain that the army quickly would make it standard therapy. But now, in 1942, more than two years later, his drugs were still not seeing widespread use.

The military delayed because until the Heydrich assassination there was still doubt about sulfa at the highest levels. Hitler himself opposed animal experiments. He knew nothing about Domagk except that the Bayer researcher had thought about accepting a Nobel Prize despite his Führer's displeasure. The physicians around Hitler spoke of increasing danger from side effects. There were stories that sulfa had killed some patients. Domagk himself investigated some of the anti-sulfa claims and found them false. Regardless, the German army hesitated, refusing to adopt sulfa for battlefield use as generally as Domagk believed they should. He worked hard for more than two years to persuade them. Much of the work was done on his own; Bayer was involved in more pressing wartime research. For many months the military medical officers ignored him, thinking that the sulfa they had was sulfa enough.

Heydrich's death led to a change in attitude at the highest levels. One result was the Ravensbrück experiments. Another was an invitation Domagk received to demonstrate his newest sulfa's abilities to a group of military medical experts in Brussels. The man who invited him in the late spring of 1942 was a Dr. Wachsmuth, a surgeon who believed that surgery was the answer to all wounds and a leading critic of the use of sulfa on the battlefield. Receiving an invitation from Wachsmuth to talk about sulfa before an audience of German army surgeons was a great step forward. When Domagk arrived in Belgium, dressed in his army captain's uniform, he was paired with a leading bacteriologist from Hamburg, a Professor Pfeissler, for the demonstrations. Together they infected a group of test rats with dirt laced with gas gangrene bacteria, then treated half of them with Marfanil. None received surgery. While they waited for results, Domagk traveled to Bruges to admire the architecture. He returned to find all the untreated rats dead, while most of those which had received Marfanil were free of disease. The experiment was a convincing success. Domagk told the military physicians that results could be further improved if Marfanil was given together with other forms of sulfa. It was clear that treating fresh wounds with sulfa could save lives. The German military hierarchy finally took Domagk's advice and started equipping every soldier and medical corpsman they could with MPE Powder—a mixture of 40 percent Marfanil, 30 percent Prontalbin (pure sulfa), and 30 percent Eleudron (a German version of sulfathiazole)—packaged in a handy sifter-top container. The rate of gas gangrene infections in German troops dropped dramatically.

The British, too, were a bit slow in adopting the widespread use of sulfa on the battlefield. Leonard Colebrook, the head of the childbed fever unit at Queen Charlotte's in London, would help change that. Soon after the Germans entered Poland, Colebrook rejoined the army as a colonel and was given the task of assessing the value of sulfa to the military. In the early days of the war, he crossed into France to oversee large tests of the drugs near the Maginot Line, coordinating efforts with Jacques Tréfouël from the Pasteur Institute, who was studying the use of the chemicals in French soldiers in the same area. Colebrook had just gotten his tests set up in the spring of 1940 when the Germans blitzed the Allied defenses. For a month Colebrook dashed around northern France trying to avoid the Nazis, collect scattered test results, and secure his lab equipment. He was evacuated from France in June along with the remains of the British forces, convinced, despite the loss of much of his hard data, that the application of sulfa to new wounds could help stop infections. He spent the next year proving the point with further tests. The wounded British soldiers returning from France in 1940 were disquietingly like the casualties in World War I: no time for quick surgery, the wounds dirty, the infection rate high. Only a small proportion of them had received sulfa, since the rout in France created conditions "unfavorable to the routine field application of a new form of treatment," in the restrained wording of the military medical reports. Colebrook's work eventually helped convince the authorities that liberal use of sulfa as early as possible was a wise idea. Along the way he also demonstrated that sulfa administration should be maintained during hospitalization as well, since many military wound infections started not on the battlefield but in the hospital, from strep carried in the noses and throats of nurses and physicians. He became an expert on the use of plastic surgery to rebuild shattered soldiers, proving along the way that local application of sulfa powder to superficial wounds could clear out strep within a few days. And he began to investigate sulfa's effectiveness on burns. World

War II was a war run on gasoline, and petrol-caused burns were killing thousands of soldiers and sailors. Large-scale burns created a huge field for infection. The burns themselves, the sloughing skin of the charred men in the hospitals, he found, could be a major source of infection, with the seeping wounds soaking the dressings and bedclothes, contaminating the cloth, dispersing bacteria into the air and into dust. Today's clean-room procedures in burn wards, with improved ventilation and dust suppression, owe much to Colebrook's work. By the time of the North Africa campaigns, the British, too, were giving sulfa to as many soldiers as possible. As a result, severe wound infections became relatively rare.

FOURNEAU AND MOST of his research group, including the Tréfouëls, were in Paris when the Germans arrived and occupied the city. A contingent of Nazi soldiers secured the Pasteur Institute. The gatekeeper there was an elderly man named Joseph Meister. Fifty-five years earlier, Meister had been famous. He was an Alsatian village boy, nine years old in the summer of 1885 when a mad dog knocked him down and savaged him. The dog was killed while it was still on top of him. When townspeople pulled the animal off they found the boy covered with slaver and blood, bitten many times on the hand, legs, and thighs. The town doctor did what he could, doused the boy's wounds with carbolic acid, and started seeking help. It seemed certain that the boy had been infected with rabies, a disease that killed almost without exception. In desperation, Meister's mother took the boy on the train to Paris, to see the great Dr. Pasteur himself. Pasteur had been investigating rabies and had been developing an antirabies serum in dogs. With little to lose and a desperate case before him, he treated the boy with his experimental vaccine. Joseph Meister recovered, and his miraculous cure added luster to Pasteur's reputation. The Alsatian boy felt a lifelong debt of gratitude. Meister spent most of the rest of his life serving Pasteur and his institute, much of it as a gatekeeper. When the Nazi soldiers arrived and demanded to be admitted to Pasteur's crypt, Meister refused to allow them in. They drew their guns and pushed him aside. Deeply shamed by the desecration,

Meister went home, pulled out his World War I service revolver, and shot himself.

Ernest Fourneau, the Pasteur researcher in whose lab the power of pure sulfa was discovered, was far more flexible. He knew the Germans. He accepted the German occupation as a fact of life. He had little to fear: He had been a leader in German-French scientific cooperation, had been a guest of the German government at the Berlin Olympics, had heard the Führer speak, and had been a guest at Göring's home. He knew many people of influence. The Pasteur Institute was kept open under German occupation, and Fourneau, like many other French researchers, continued to run his laboratories.

By 1943, almost every German, American, and British soldier either was carrying sulfa or could get it quickly from the nearest medic. Every American soldier was given a packet or two of powdered sulfa to carry in his first-aid kit, packaged so it could be used with one hand, with instructions to sprinkle it on wounds immediately. Battlefield medical aides doled out powder and pills to anyone threatened with any sort of infection. The British had May & Baker 693, and the Germans their sifter-top canisters of MPE Powder. Sulfa production on both sides of the Atlantic was increased year after year until it was made by the hundreds of tons. It was still not enough to meet the wartime needs.

Russian soldiers did not have as much access to sulfa, and in Japan a lack of sulfa might have helped tip the scales of war toward the Americans. Good numbers are hard to find, but several anecdotal reports indicate that American forces had far better access to sulfa, which meant better treatment and faster recovery from wounds and dysentery, which meant an edge in battle-readiness. The Japanese were so hungry for sulfa that one U.S. soldier in a Japanese POW camp remembered guards "willing to do almost anything" to get at the few sulfa tablets the GI prisoners received from the Red Cross. The Americans turned it into a small business, with prisoners fashioning exact duplicates of their sulfa pills out of plaster of paris and trading them to the Japanese guards for tobacco, food, and favors. "Why they never wised up I'll never know," a prisoner wrote, "but I guess some of them must have wondered why the American wonder drug would not work on his ailment."

U.S. military physicians in the South Pacific were seeing fewer infections and faster healing of even the deepest wounds if they doused the area immediately with sulfa and gave it by mouth as a follow-up. Guerrilla fighters in the Philippines did not get much in the way of medical supplies in the early 1940s, but when submarines did manage to smuggle in a shipment, the medical boxes always contained three things: bandages, quinine for malaria, and as much sulfa as possible.

Almost every GI wounded in action remembers something about sulfa. So does everyone who provided health care. Wheeler Lipes, a twenty-two-year-old pharmacist's mate on the USS *Seadragon,* a submarine on patrol in the South China Sea, was the only medical helper on board when a young sailor reported a terrible pain in his gut. He thought he had constipation, and he asked Lipes for a laxative. It was September 8, 1942, the sailor's nineteenth birthday. Lipes was no doctor, but he knew the difference between constipation and an inflamed appendix. He put the sailor in a bunk and monitored his condition. Within hours the sailor's temperature started rising. Lipes figured his appendix might be about to rupture. If that happened, the birthday boy would die. Lipes was pretty sure of his diagnosis. He also knew that the only thing that could be done was surgery. The sub was days from any friendly port.

Lipes reported what he knew to the sub's commander. The skipper checked on the patient's condition. Then he asked Lipes what he was going to do. The pharmacist's mate replied, "Nothing." The *Seadragon* was not equipped with surgical equipment, did not have proper anesthesia or any way to monitor blood pressure, and, most important, lacked a surgeon. Lipes had assisted with a few operations Stateside, but he had never made an incision. Surgery was unthinkable. The skipper asked the pharmacist's mate if he thought he could take out the appendix. Lipes said he thought maybe he could, but that the chances were good that the patient would not pull through, and, given his choice, he would rather not try. The skipper said, "I'm out here every day, and my job is to sink enemy ships. I fire torpedoes and sometimes I miss. But I keep at it because that's my job."

"Sir, I can't fire this torpedo and miss," Lipes replied.

The skipper ordered him to take out the seaman's appendix. Lipes, "highly motivated," as later reports said, asked a couple of young officers to help him. He cleared off a wardroom mess table to serve as an operating table, lined a tea strainer with gauze to make an anesthesia mask, and bent the handles on teaspoons to use as retractors. He used "torpedo juice"—a form of alcohol used to fuel the sub's weapons—to sterilize some rubber gloves and a pair of pajamas he wore in place of a surgical gown. He ground some sulfa tablets into a powder. He did not have a complete scalpel, but he had a scalpel blade, so he boiled that and anything else he thought be might be able to use for instruments.

The skipper took the sub down to 120 feet, where the water was calm enough to give the makeshift medical team a stable base. Lipes was not supposed to have ether on board, but he had bought some at their last port and hidden it away; now he told a young officer how to drip the ether onto gauze in the tea strainer and then showed him how and how often to place it over the patient's mouth and nose. When the patient was unconscious, Lipes made a three-inch incision over the spot where he figured the appendix was. He pulled back the skin and explored with his fingers. He could not find it. "When I got to the appendix, it wasn't there," was how he remembered it. He thought for a moment that the sailor's organs might be reversed; he had read about that, people whose hearts were on the right side instead of the left. Then he slipped a finger under the coecum—the blind gut—and found what he was looking for. The appendix was grossly swollen and looked two-thirds gangrenous. One end was black. It was adhering to the lining of the intestines in three places. It would have to be cut loose in order to be removed, and if the gut lining were punctured in the process, it would release a flood of bacteria into the abdomen. "What luck," Lipes thought. "My first one couldn't be easy." But he had no choice. He painstakingly pared the appendix from the intestines. He did not think that he had nicked anything important. Then he removed the diseased organ, tied off the stump, and cauterized it with Lister's old standby, carbolic acid, tempered with a bit of torpedo juice. He sprinkled sulfa powder into the cavity and used more on each layer of tissue during closing. The operation took two and a half hours.

To almost everyone's surprise, it worked. A few days later, his patient was back on active duty. The ship's cook told Lipes, "Doc, you must have sewed him up with rubber bands, the way he eats."

SULFA'S IMPACT WENT far beyond treating wound infections. Dysentery was a serious problem in wartime, especially in warmer climates. The disease could be caused by several types of protozoa and bacteria, but the worst was caused by *Shigella* bacteria. Sulfaguanidine, one of the newer forms of sulfa drugs, was found to be especially effective against *Shigella.* Sulfaguanidine's advantage over other sulfa drugs was that it stayed in the intestines longer, giving it more time to work against intestinal diseases like dysentery. Serious outbreaks of dysentery in the Guadalcanal area were corralled with sulfaguanidine, saving countless lives. In an epidemic among Allied troops in New Guinea, the drug was found to be "dramatically effective"—only two deaths among ten thousand patients treated. Treated men recovered faster, and that meant more troops for the front lines. Sulfaguanidine, in the opinion of at least one historian of medicine, "revolutionized" the treatment of dysentery during the war. In the Pacific theater, sulfa drugs—a medicine discovered in Nazi Germany—helped the Allies win the war.

The lesson was not lost on military planners. After the first few months of war, the U.S. Army decided it wanted to learn everything possible about sulfa's use against other diseases. Meningitis, for instance, was a serious military problem, capable of causing fatal epidemics almost anywhere large numbers of soldiers were closely quartered. Today university dormitories are occasionally hit with an outbreak. In World War II, the worst place to be was a barracks in an army training camp—no matter what nation the army served. Seven months after Pearl Harbor, there was an explosive outbreak of meningococcal meningitis among U.S. troops in several training camps, the rate of infection doubling, then almost tripling among new draftees. It reached a peak in the winter of 1942 with a severe outbreak in several camps in the Southeast.

Military physicians knew that sulfa could stop meningitis after it started. Now they wanted to see if it could *prevent* the disease. They worked from the observation that many people carried potentially fatal meningococcus in their throats without developing the disease, just as Colebrook had found many nurses and doctors acting as carriers for strep. Early small-scale tests indicated that finding meningitis carriers and treating them with sulfa could eliminate the meningococci, lowering the carrier rate. What was not known was whether this in turn would lead to a decline in the rate of infection. Research had been crippled because of logistics: To know the answer for certain, researchers needed to run the experiment on two large groups at the same time, groups similar in every way except that one would receive the sulfa and the other would not. Those kinds of tests on a massive scale had never been possible. Most drug testing had been done on individuals —physicians or their family members—or on tribal groups, or prisoners, or, like Colebrook's research, on patients who were given the drug and then compared to historical data. The general populace was never a good place to test. People were uncontrollable; they insisted on traveling where they wanted, meeting whom they liked, and forgetting on occasion to take their pills. Prisons seemed promising but were less than perfect: The numbers of subjects was relatively small, prison populations tended to mix, and prisoners were not generally renowned for their compliance. Soldiers in training camps, on the other hand, were already under full control when it came to housing and diet, tended to obey orders, and could be overseen twenty-four hours a day. They were perfect test animals. So with an epidemic of meningitis beginning to rage in the Southeast's army bases, U.S. military leaders decided to use training camps for the largest mass drug tests in medical history.

The first "laboratory" was an army base in rural Mississippi. The first guinea pigs were eight thousand soldiers, every man in one area of the huge base, who were given a low dose of one sulfadiazine pill with every meal. A commanding officer handed each soldier a pill as he entered the mess hall. A sergeant made sure every soldier swallowed it. About nine thousand "controls" in another part of the base received no drugs. After six months researchers found that twenty-three of the controls caught meningitis. None of the treated men got the disease. The experiment was expanded to another camp, a total of fifteen thousand experimental subjects and nineteen thousand controls. The meningitis carrier rate fell dramatically; in one part of the base, the drug cleared the bacteria from every carrier. During the same period, the carrier rate rose in the controls. Among the fifteen thousand treated men, only two developed meningitis during a two-month period; there were forty cases among the controls. The navy started its own set of tests, as did the air corps. By the end of 1943, after six massive tests run by the military, the results were clear: Mass prophylaxis with sulfa could contribute to a significant decline in the rate of meningitis in the armed forces. It was certain that the drug was effective in controlling specific outbreaks.

During World War I, meningococcal meningitis had killed as many as two-thirds of its victims. When a wave of meningitis hit British military bases early in the Second World War, physicians using sulfa—mostly local favorite M&B 693—cut the fatality rate by more than half, dropping it below 15 percent. U.S. physicians using sulfadiazine got even better results, knocking the fatality rate below 10 percent, finding that sulfadiazine even in massive doses caused an "astounding" lack of fatal side effects. By the end of the war, using increasingly refined treatment protocols, the military would drop meningitis fatality rates below 4 percent.

The ability to give drugs to thousands of compliant soldiers, the hierarchy needed to effectively gather and analyze results, and the range of diseases and wounds being treated made the military an extraordinary test laboratory for sulfa. The results were convincing. Acute respiratory diseases, including influenza, pneumonia, bronchitis, and other diseases, had killed almost 50,000 U.S. soldiers in World War I; during World War II, with twice as many men and women in uniform, only 1,265 died. The main difference, according to the official U.S. military medical report on the war, was the wide use of sulfa drugs.

Sulfa was best known for beating the deadliest diseases, but in the early days of the war it probably meant more to soldiers for another reason, another disease, one not often fatal but always unwelcome: gonorrhea. One physician called it the "wet-diapered and puny little bastard that would neither get well nor die." It had afflicted humankind for centuries. One of the most famous victims, John Boswell, the biographer of the great English literary figure Samuel Johnson, suffered nineteen bouts of gonorrhea, a frequency reflective of both his personal conduct and the lack of decent condoms. In the late eighteenth century, Boswell's day, the only condoms available were made of either linen (which had to be wetted first but had the advantage of being launderable) or sheep's or goat's gut, pickled, often scented, and secured in place with ribbons. Ribbons in regimental colors were particularly popular. Despite—or perhaps because of—this, gonorrhea was everywhere. The disease was little more than embarrassing and uncomfortable at the beginning but, left untreated, could progress to much more serious conditions, resulting in sterility, arthritis, even death.

The treatment in the old days could be more painful than the disease. Gonorrhea often made it difficult to urinate, because it could cause a narrowing of the urethra carrying urine from the bladder. The usual treatment for men in the eighteenth century involved passing a curved metal rod up the penis. When treated in this way, Boswell fainted.

By the early 1930s, although there was still no cure, things had changed for the better. A number of ointments were available to ease discomfort. Silver picrate ointment actually did some good fighting the bacterium that caused gonorrhea, although it could be agonizingly irritating. One of the most popular treatments was "fever therapy," in which the patient's temperature was raised artificially (through mechanical means or via a shot of a vaccine), the raised temperature helping to discourage the disease. Early sulfa drugs, including pure sulfa, had variable results on gonorrhea but cured it often enough to lead to an orgy of self-medication with over-the-counter sulfa in the mid-1930s. The next generation of sulfa drugs was much more effective. M&B 693 was especially good and during World War II became the drug of choice for British soldiers and sailors. Four or five days of pills usually cleared up all symptoms, an amazing improvement over any previous treatment, yielding complete cures that "sounded like a fairy story to venerologists," according to one historian. It also led many victims to stop taking the drug too quickly, when the symptoms of gonorrhea were gone but enough bacteria remained in the body to pass the disease to others or reinfect the patient. Those taking the drug thought they were cured and resumed their sexual practices, sometimes selling the leftover sulfa or giving it to others, and the pattern would repeat. The ultimate victims were often unwitting wives or husbands whose partners gave them the clap. Children suffered as well; gonorrhea in the eyes of a newborn could blind the infant, one reason that antibacterial drops are immediately put in the eyes of newborn babies. Regardless, mass use of sulfa against gonorrhea was undertaken by the public well before any large-scale testing was done.

Researchers hurried to catch up. Early experiments were done on prison "volunteers" whose sexual activity in particular, it was thought, would be easy to monitor. As it turned out, however, the tests were inadequate—prisoners seemed able to do a number of things that were more covert than scientists had hoped.

So researchers turned again to the armed forces. Here the clap was considered a significant issue of military preparedness. During World War I, when it took two or three months to clear up a case, gonorrhea had been second only to the flu as a cause of disability and absence from duty. Sulfa promised to cut the full course of treatment to three weeks, getting hundreds of thousands of men off the disabled list and back to the front lines faster.

The first large-scale military test was undertaken with fourteen hundred GIs, who were given sulfathiazole as a preventive medicine, a way, researchers hoped, to stop gonorrhea before it could get started, an attempt to "Stamp Out Gonorrhea Now!" as the army circulars encouraged. The military physicians succeeded in significantly lowering the incidence of the disease in the sulfa-taking group compared to the four thousand or so soldiers in the control group, men who had the same habits but not the medicine. Sulfa was prescribed in vast quantities. Some posts began handing out pills to every man who went to town in the evening and again when he returned. Hundreds of thousands of men were cured.

ON DECEMBER 11, 1943, Winston Churchill boarded a plane in Cairo. He was flying to Tunis, where he was planning a few days' rest at Dwight Eisenhower's villa, presciently nicknamed "the White House," near the ruins of Carthage. Churchill was nearing the end of an exhausting trip. He had conferred with Chiang Kai-shek in Cairo, then flown over Palestine, Iraq, and much of Iran to meet with Stalin and FDR in Teheran. The Allies were getting ready to take France back from the Germans, planning for D-Day was under way, and the three-day meeting of the Big Three Allied leaders firmed up arrangements with all the players. Then Churchill flew back to Cairo for further talks with FDR. The British leader, sixty-nine years old, a heavy smoker and legendary drinker, was overworked, overweight, and mentally and physically worn out by the time he got on the plane for Tunis. His flight was delayed along the way, and he spent an hour sitting on his luggage in a cold wind in some godforsaken airfield in the desert. When he finally made it to Eisenhower's villa, his throat was sore. The next day his temperature was 101. His personal physician, Charles McMoran Wilson, Lord Moran of Manton, who had been at his side through the entire trip, could not immediately determine the cause. Worse, they were, medically speaking, in the middle of nowhere. Lord Moran did not know the local physicians, the hospitals were not well equipped, and medicines and equipment were meager. There was no nearby laboratory to help him determine the cause of the illness. He did not even know where to get milk to boost Churchill's diet. So he sent to Cairo for equipment, two nurses, and Dr. Evan Bedford, brigadier in the Royal Army Medical Corps and consulting physician to the British forces in the Middle East and North Africa. The Cairo team arrived quickly, took blood, and found that Churchill's cell counts seemed normal. But the patient was weakening. Lord Moran rushed in a portable X-ray machine from Tunis. The X-rays revealed the source of the problem: There was a shadow on the left lung.

Churchill had pneumonia. "It means we can begin giving him M&B straight away," Lord Moran wrote in his diary. By the time he did, Churchill's heart had begun giving out. He suffered two bouts of atrial fibrillation and an episode of cardiac failure. Lord Moran treated him with digitalis and sent urgent calls to every leading specialist in the Mediterranean: a heart man from Cairo, a sulfa expert from Italy. Churchill's lungs became more congested. Lord Moran called in the prime minister's family members. With the Allied effort nearing a crisis point, it appeared that Great Britain's war leader was dying.

Lord Moran canceled a planned trip to Italy and conferred with his growing medical team about what, if anything, they should do next. Then, finally, the sulfa kicked in. Churchill's temperature returned to normal. He began to take food. And he began, impatiently, to take the course of convalescence. By Christmas he was actively participating in the planning of the Allied landing at Anzio, going on short walks in his robe and slippers, and telling visitors that the entire affair had been "touch and go." Two weeks after he first fell ill, he flew to Marrakesh and then home.

Everyone was convinced that he owed his recovery to the new medicine. When he returned to England, Churchill joked with reporters, telling them that he was now referring to his physicians, Moran and Bedford, as "M&B." He joked that he had been happy to learn that he could take the new medicine with a shot of brandy, which increased its attractiveness. Then he told them, seriously, "This admirable M&B, from which I did not suffer any inconvenience, was used at the earliest moment; and after a week's fever the intruders were repulsed. I hope all our battles will be equally well conducted. . . . There is no doubt that pneumonia is a very different illness from what it was before this marvelous drug was discovered."

It was sulfa's last great moment.

CHAPTER TWENTY-TWO

Gerhard Domagk spent most of 1942 and 1943 traveling the European theater of war, speaking to German army physicians about sulfa's benefits, outlining the most effective methods of application, and promoting the medicine's widespread use. In Hungary he toured medical facilities. In Milan he enjoyed the opera at La Scala with a contingent of Nazi officers. His new forms of sulfa finally were being widely used, as he had hoped, to solve the problem of wound infections, but still he was deeply unhappy. He had never fully recovered from his imprisonment. He found it difficult to concentrate, had trouble sleeping, and continued to suffer from occasional chest pains. "I was still very much run down," he wrote of these years. "I had not recovered normal health since the shock of my imprisonment. . . . Under such pressures, the quality of my research was dramatically reduced." Still, much remained to be done. He was now intent on curing tuberculosis, which had so far resisted sulfa treatment. But Bayer, increasingly involved in making medicines for German soldiers, was decreasingly able to support Domagk's drug-development efforts. Tuberculosis was a disease of the poor and weak, not German soldiers. It was not important to the Nazis. Funding was limited, scientific communication was hindered, and many of Domagk's lab assistants were being called into the army.

By 1942, the war was beginning to go badly for Germany. On the home front, Germans were starting to suffer shortages of food, gasoline, and clothing. Domagk had trouble finding shoes for his children. Then the bombing started in earnest. As early as 1940, Great Britain's Royal Air Force had begun experimenting with daylight raids on German industrial targets—Domagk was compelled once that summer to take shelter in a cellar—but the British lost too many planes, and the raids were suspended. They restarted in 1942 in western Germany—especially in the Ruhr, the industrial area that included the Bayer plants at Elberfeld and Leverkusen, and the town of Wuppertal, where the Domagks lived. On May 30, the British launched a "thousand-plane raid" against Cologne, thirty miles south of Domagk's home, flattening the city, leaving only a few hundred houses and the central cathedral standing. Two nights later the RAF attacked Essen, fifteen miles to the north of Domagk's home. Thousands of civilians were killed. The raids seemed designed to spread terror as much as destruction.

At the end of May 1943, the RAF launched its first major raid against Wuppertal itself. In less than an hour, 719 planes dropped nineteen hundred tons of bombs on the small city, killed 2,450 civilians, left more than 100,000 homeless, started a firestorm that burned more than a thousand acres, and destroyed more than half the town's buildings. Domagk, who was away giving lectures, heard via a telegram from his wife that their home and family were safe. He rushed back. A few nights later, the Allied bombers came again. He woke the next morning to find the sunlight "dimmed with yellow, red, white thick fog and everywhere soot and ash." The Allies had destroyed most of Düsseldorf, twenty miles away; Domagk was seeing the aftermath carried on the wind. The bombings continued. During this period Domagk, sitting at his desk in an almost empty lab, wrote:

> Of sixty-five employees here in my laboratory, there are hardly ten left. Many are dead. Others have been made homeless and are now desperately searching for a roof over their heads. Some are having difficulty in getting to work because of lack of transport—there are still a great many fires burning and many roads are not passable. Tramcars and railways are not running at the moment.

Fortunately our lives have been spared yet again. I didn't even hear the fire alarm—I had gone to bed late and utterly exhausted, but when I woke in the morning the whole town was ablaze. Above us roared the all-too-familiar cacaphony of attack, the bombs and antiaircraft guns and a deafening noise. Minutes later the whole sky over us, from Elberfeld to Cronenberg and Remscheid, was a burning bloodred. As soon as the shooting stopped, I went to where help was most needed. The factory had escaped serious damage. It had been hit by a few incendiary devices that were soon put out, so I decided to go to the hospital to see if I could help. But the road was already closed. Together with my laboratory assistant, I tried to get there by an alternate route, behind the power station across Schiller Square. Behind the power station, we crept along the rubble-covered streets, under a large chimney that was still standing. Most of the fires had been put out. The houses in Schiller Square were still burning; in front of them sat the people who lived in them, with the few belongings they had been able to save. People were throwing mattresses, blankets, and clothes out of the windows of burning houses. My assistant and I made our way through the smoke and sparks and between burning falling wooden beams. The hospital itself was ablaze. Block 1, where Dr. Gehrt had his children's clinic, had quickly been cleared, as had Block 2, the surgical clinic. In Block 3, the roof rafters were ablaze. Some of the buildings were preserved. All the sick people had been successfully evacuated. Professor Klee had been able to save some of the medical records and books. I came back through the subway under the rail track to the tarmac road—that was the only way through. In the subway, people were sitting about with beds and clothes in front of them. Outside they were surrounded by flaming buildings and debris. I am taking a family with two children into my house—the mother couldn't move any further. The area of Steinbeck around the station was decimated, as was the main railway station. Many of my colleagues are completely destitute. Klarer's flat was badly damaged. On Friday morning, we effected some essential repair work on the factory and then went to the hospital to help with the injured. All the shrapnel wounds

> were treated with Marfanil-Prontalbin. In this attack, the English started up at exactly the same spot where they left off last time. . . . I am very glad that Gertrude and the children are not here.

After the first attack on their town, Domagk sent his wife and older children to their summer home on the Baltic. The seaside summer-house had been upgraded; electricity and running water had been installed. It was adequate, and it was in a relatively deserted area, safe from British bombers. The two youngest, Hildegarde and Jörg, were sent to Brandenstein, where a friend of Domagk's, the Baroness von Breitenbuch, looked after them. Domagk stayed in Elberfeld to try to keep his research going; except for occasional visits, he spent the rest of the war separated from his family. His laboratory, like almost all the rest of the Bayer factory at Elberfeld, was still standing. It appeared that the Allies were sparing the factory on purpose. Perhaps they had plans for it after the war.

Domagk loathed the British fliers for what he considered their senseless, terrorist air raids. He believed, as many Germans did, that the bombing was directed more at civilians than at military targets. Neither was he enthusiastic about the Nazis, especially after his experience with the Gestapo. He believed that Hitler had come to power because 6 million Germans had been out of work and were looking for anyone who would promise them a future. Domagk never joined the National Socialist Party as Hörlein had, he refused to drape a swastika on his porch during state holidays as many of his neighbors did, and he rarely signed his correspondence "Heil Hitler" as almost all of his professional colleagues were doing. "The National Socialist system started with lies and ended in cruelty and blood," he later wrote. War to him was the ultimate madness. He sent his family to safety and lowered his head to his microscope.

Work was the only thing that brought any light into his life. Despite everything, he kept working to find new cures. He sensed the end of the war coming now and expected to see the same scenes he had seen toward the end of World War I in Flanders: crowds of refugees, smashed cities, shattered health systems, bad water, no food, and rampant disease. It was a perfect setting for epidemics, especially tuberculosis, a disease that preyed on the weak and flourished where there was poverty. It was a horrible disease, capable of turning healthy lung tissue into a germ-riddled semisolid, its victims literally coughing their lungs out. Domagk threw himself into a search for a medicine that would stop it.

TB was a difficult disease to study. The germs were encased in a waxy coat that seemed somehow to protect them from the body's defenses and, researchers thought, might help armor them against drugs. They were very slow-growing, and animal tests took a long time to run. They were also very dangerous, highly contagious, perfectly capable of accidentally infecting and killing a careless laboratory assistant.

The only sulfa variation that had ever shown any significant effect against tuberculosis in animal tests was sulfathiazole, but it was not strong enough to use in humans. Domagk wanted new drug variations, but those, too, were getting harder to come by. Klarer and Mietzsch lost the remainder of their enthusiasm for working with Domagk in 1942 after he offended them yet again, they believed, by demeaning their contributions in an article in a German medical news journal. Domagk, stung after being reminded that on the Prontosil patent it was the chemists' names that appeared, not his, wrote his feelings in his journal. "That is not the point at all. Helping is what matters, nothing else. Patents, patents. I have never claimed for myself the chemical development of the sulfonamide substances. What the chemists were expecting when they developed them, I do not know. . . . Who has actually achieved the better work, that I do not dare to decide." The chemists complained again to Hörlein, who understood that having their names on the patent for Prontosil was very important in ways that went beyond mere credit. Chemists who had their names on a Bayer patent received a small share of the proceeds generated by the patent; in Prontosil's case that represented a significant sum of money. The practice was central to rewarding the most imaginative, money-making workers. Mietzsch and Klarer earned money directly from the sale of Prontosil. Domagk did not. Instead he received the lion's share of public credit. That seemed fair enough.

But the publication set off yet another string of heated arguments within Bayer. This time the pressures and privations of the war led to frayed tempers. Hörlein threatened at one point not to renew Domagk's contract. Domagk replied that as he was getting little support for his TB research in any case, he would rejoin the army full-time. Hörlein eased off, smoothed over his team's personal feelings, and once again set the ship right. One result was a bit more money for TB research. Another was the end of Mietzsch and Klarer's providing chemicals for the effort. The chemist now working with Domagk was the group's "new" man, Robert Behnisch. It turned out well. Just as Fourneau's French group had broken Prontosil into two fragments and found that the sulfa side chain had all the power, so Behnisch decided to break the sulfathiazole molecule into two parts, the sulfa and the thiozole. Then he put the sulfa portion aside and started working with the thiozole portion alone, cracking open its central ring structure at various points, moving atoms around, adding and subtracting side chains. There were enough small effects, enough tantalizingly positive results, to keep him going. Working between air raids in 1943, Behnisch began coming up with increasingly effective variations. Opening the thiozole ring at one particular point gave the best results, and led to the birth of a new family of molecules called "thiosemicarbazones." Working through 1943 with a skeletal staff, frequent interruptions, test animals in short supply, windows blown out, the library burned, Domagk and Behnisch began making exciting progress. Perhaps, despite everything, there would be a cure for TB.

By 1944, Domagk could no longer suppress his feelings about the Nazis. Everyone around him was in rags except for party officials. His diary entries included barbed descriptions of local bureaucrats striding through the rubble: a baker "strutting along like an Italian cockerel in his brown gala uniform," Gestapo men "crawling around" in the wreckage "sticking their noses into things," the former central-heating repairman now a Nazi official with "trousers made of the finest black cloth, brown coat and brown raincoat. Every time I saw him he appeared more magnificent." When the official greeted Domagk with a hearty "Heil Hitler!" the scientist refused to respond.

Domagk's personal feelings were somewhat at odds with his professional life. He might have been working for the good of humanity, but he was also working for a firm that was fully engaged in furthering Nazi wartime aims. IG Farben became the most notorious business in the world as more was learned about its misuse of concentration camp inmates to produce synthetic gasoline and rubber. A Farben subsidiary produced the Zyklon B poison gas used to kill Jews at Auschwitz. By 1944, almost one quarter of the workers at Duisberg's huge Leverkusen factory were forced laborers, mostly Poles and Russians brought in to replace the war's labor losses. "Foreign workers," as they were euphemistically called, were also employed by the hundreds at Elberfeld, where Domagk must have seen them shuffling to their jobs. Some things he could turn his head away from. Some he could not.

He undoubtedly knew, for instance, that his friend Philipp Klee, the head of internal medicine at the Wuppertal hospital, the man who oversaw some of the first human tests on Prontosil, had managed for years to protect his Jewish wife, Flora, from the SS. The Klees, husband and wife, had endured intimidation and insults. The physician had lost patients and much-needed funding because of his marriage. But by 1944, Klee found it impossible to shield Flora any longer. One of the last Jews in Wuppertal, she was arrested in November and deported to the Theresienstadt camp in Czechoslovakia. Klee, desperate, with Germany crumbling around him, set off on a thousand-kilometer trek to get his wife back. By the time he reached Theresienstadt in early 1945, however, the camp had already been liberated and the area was in chaos, records sketchy, and roads packed with displaced persons. He could not find Flora anywhere. There was no way of knowing if she was dead or alive. With no other options, Klee returned to Wuppertal. And there she was, waiting for him. Flora, heavily disguised and traveling with a caravan of refugee Gypsies, had found her own way back.

Domagk's family was still scattered. His wife was at their summer home on the Baltic; the two youngest children were still with the Baroness von Breitenbuch; his eldest son, Götz, was in the army learning how to fire antiaircraft guns; his second son, Wolfgang, thirteen years old when the bombs started hitting Wuppertal, was at a boarding school in Thüringen. In March 1945, in the last convulsive weeks of the war, Domagk was horrified to hear that Wolf and his classmates were given rifles, marched through the night, and ordered to attack an advancing Russian army. This news was the last the Domagks had heard from him.

Domagk did know, however, about the fate of his mother. His father had died some years before the war, but his mother lived through most of it with his sister Charlotte in their family home in far eastern Germany. After the Russians occupied the area in mid-1944, Domagk heard no news for six months. Then he heard from Charlotte. She and their mother had been moved out of their house by the Russians, then allowed to return. His mother had later been sent to a home for the elderly, then was ordered out of that as well. The German town in which he had been raised had been declared part of Poland. Domagk's mother and sister were evicted from their home and ended up traveling on foot with a refugee group. His mother was too old to walk far. Exhausted, she was put into a cart. Her coat and jewelry were stolen. They reached a refugee camp on the side of a river, Charlotte was not certain where. And there his mother had starved to death. Charlotte made her way to Wuppertal, to live with Domagk and his family.

By the summer of 1944, all work had ceased at the Bayer labs. They were out of chemicals, out of animals, and out of staff, almost all of whom were lost, wounded, or dead. In the spring of 1945, the Allies occupied what was left of Wuppertal and took charge of the Elberfeld plant. The British control officer for the Bayer facility, who worked for the rival May & Baker firm in England, "was constantly at the factory, studying and copying all the records, all the chemical formulas," a German researcher remembered. Domagk's home had survived the bombing and was turned into a billet for soldiers. Most of the large Elberfeld plant along the Wupper River, including Domagk's lab building, was still intact. The giant Bayer facility at Leverkusen, however, did not fare as well, with about a quarter of Duisberg's dream factory destroyed before the advancing Americans handed over control to the British, who began supervising the repair of the plant. One of the first products made at British-occupied Leverkusen was a batch of Prontosil.

By Christmas 1945, the Domagk family except for Wolf was together once more. The British had restarted work at Elberfeld, and Domagk was again working on his cure for tuberculosis. His wife, Gertrude, retreated to their summer home on the Baltic, where the family was set to gather for the holiday. And there, one day toward Christmas, they saw a young man walking up the road toward the house. It was Wolf. He had somehow survived the student attack on the Russians, lived through the last days of the war, evaded enemy soldiers, and trekked across the entire country, a teenager traveling alone. After eight months of wandering through the ruins of postwar Germany, full of stories, apparently healthy, he joined his family for what turned out to be their best Christmas since the start of the war.

GERHARD DOMAGK'S FATE was in this way happier than that of Ernest Fourneau and Heinrich Hörlein, the two competing laboratory managers who had clashed over the development of Prontosil.

Fourneau, who had spent the war getting along with the Nazis and continuing his laboratory research, received a letter from the Prefecture of Police of Paris soon after the city was liberated in 1944. He had been denounced as a collaborator, the letter said, and he was requested to present himself at the nearest police station so that he could be placed under house arrest. Fourneau immediately went into hiding. "I knew that a certain group, quick to judge on appearances, would denounce Mr. Fourneau, among many others," his right hand in the lab, Jacques Tréfouël, wrote later. "I decided that my Master was blameless and . . . I convinced him—with great difficulty—not to remain at his home. I had informed the French military authorities that I knew where he was staying after he left with the help of two faithful colleagues from the lab, and I told them that he was ready to give them, through me, any information they might want from him." Fourneau remained hidden for two months, negotiating terms, before giving himself up for arrest. He was questioned at length, an exercise that provided him with the opportunity to defend himself against a miscellany of vague charges: that he had shared scientific secrets with the Germans (he argued that his research was humanitarian and he shared it with all nations); that he had relations with German women during the war (yes, he had hosted two young German house guests, but was all innocent enough, he explained: One, a delightful young woman who provided intelligent company for his wife, was the daughter of a family he knew in Germany; the other was an invalid recovering from an auto accident); that he collaborated with the Nazis (most of the French populace, he replied, recognized the German-approved Pétain government as the official French government, and he, like the majority, had no reason not to follow that government's orders—orders which, his questioners might remember, officially recommended collaboration). Yes, he went to some receptions at the German embassy, and concerts and plays as well, always paying out of his own pocket. No, he was not anti-Semitic. "I am more interested in social changes than political ones," he told the police. "What is funny is that the German industrialists who work in my field consider me their main adversary. At the beginning of the war, a brochure was published that accused me of working in Germany to steal the secrets of German science, and of having copied the most famous German products."

That was true. It was unlikely that Fourneau collaborated any more than did hundreds of thousands of other Frenchmen during the occupation. He was held in custody for three months while many of his colleagues agitated for his release, telling the authorities that Fourneau was not only a determined competitor to the Germans but that his expertise was needed to advance French science. He was finally released without being formally charged. His career, however, was over. He lost his directorship at the Pasteur Institute and spent the last years of his life working as an independent researcher, coming up with little of importance.

Heinrich Hörlein, too, was arrested, imprisoned by the Allied authorities on August 16, 1945. The charges against him were if anything more vague than those made against Fourneau. The Allies had rounded up a number of executives from IG Farben and held them while they gathered evidence of corporate complicity with the Nazi regime, ranging from the production of war matériel to the use of forced labor, from the seizing of foreign factories to the production of poison gas. A great deal of effort went into detailing the deaths of thousands of concentration camp workers who were enslaved at two

Farben production plants set up near Auschwitz. Two years would pass before Hörlein was put on trial along with twenty-three other business executives, defendants in what was popularly called "The Farben Trial" at Nuremberg. Hörlein, like all the defendants, was formally charged with the commission of crimes against peace through the planning, preparation, initiation, and waging of wars of aggression and invasions of other countries; war crimes and crimes against humanity through participation in the plunder of public and private property in occupied countries; participation in enslavement and forced labor; and participation in a conspiracy to commit crimes against peace. The charges were open-ended and purposefully so, a door left wide open to allow free-ranging questioning, a net designed to catch fish of all sizes. The vague indictment also hinted that the government prosecutors had a relatively weak case—this was the sixth of twelve trials in the second round of the Nuremberg Trials; all the biggest Nazi players had been tried, and the most sensational cases—like the Doctors' Trial recounting the horrors at Ravensbrück—had been completed. Now the Allies were working through a variety of less important cases that attracted waning public interest. If nothing else, the Farben trial attracted the attention of top business executives around the world as well as their lawyers. Many of them ran companies that also had helped their governments wage war. If German businessmen could be tried for crimes against humanity, could not—had things gone differently—the men who ran, for instance, Ford, Lockheed, or DuPont? It was a chilling thought.

Hörlein in the dock looked anything but a master criminal, instead appearing a paunchy and mousy-looking manager (he was sixty-five years old when the trial started), more bewildered than threatening. He wondered aloud how a simple businessman like himself could get into a fix like this. The trial's prosecutors, however, had a different view. To them Hörlein was a mastermind, a crafty actor with "a fox's face," a brilliant adversary clever enough to dodge every question and skilled enough to fool the judges. He might play the part of a mild-mannered bumbler, fussing with his papers for minutes at a time while looking for the answers to questions (the prosecutors believed that this was a trick as well, a way for Hörlein to gather his thoughts during cross-examination), but beneath it all he was simply evading justice. "I was very fortunate to have found an ideal place in Elberfeld for my life's content—to work in synthetic pharmaceutical research and to help suffering humanity," Hörlein told the court. The prosecutors for their part believed that he caused more suffering than he eased. They zeroed in particularly on one incident in which methylene blue produced at Bayer was used in human experiments against typhus at Auschwitz. They spent hours grilling Hörlein, trying to tie him directly to the human tests. They were never able to prove it. If there was any documentation, it was gone. His association with the shipments, if there was any, was so tenuous, made through so many layers of executive command, that the prosecutors could not establish it clearly. "The Professor," as the prosecutors called Hörlein, tied them in knots. "Hörlein demanded that the validity of each question be proved by a document," wrote the deputy chief prosecutor. "Professor Hörlein saw to the very end of the line, answering later questions before they were asked, explaining unresponsive answers to past questions, questioning the prosecution himself." He rarely gave a simple response, shuffled his papers, and at every turn assured the judges that his only interest was in answering questions as precisely as possible. He was polite and helpful, apparently trying his best to guide the tribunal through Farben's tangled management structure, appearing more meticulous than misleading. He gave the impression that if he had erred at all, it came from following without question the orders of his government and upper management. If this was the case, then thousands of other German executives—as well as executives in many other nations—were guilty of the same thing. The court could not afford to travel down that path. Still the prosecutors kept hammering away, keeping Hörlein on the stand for two days, asking hundreds of questions. They came up with nothing. Finally one of the presiding judges blew up, blamed the prosecution for wasting the court's time, and excused Hörlein from further questioning.

When the trial was over Hörlein was acquitted of all charges, as were almost half of the other defendants. The remaining thirteen Farben executives were given short jail terms, with credit for time served. Most were soon back to work. The Allies, after going through all of IG

Farben's files, commandeering its patents, and studying its production processes—Domagk said the laboratories at Leverkusen and Elberfeld were turned into "a free information center for passing foreign chemists"—finally broke the giant cartel into pieces, dismembering it into its constituent companies. One of them was Bayer, which rose from the wreckage of the war, still strong and still inventive. Within a few years, it was again one of the world's leading pharmaceutical and chemical companies.

Heinrich Hörlein, after an appropriate amount of time out of the spotlight, was eventually promoted to chairman of the company's supervisory board.

CHAPTER TWENTY-THREE

ON A FALL AFTERNOON in 1947, Gerhard Domagk returned home exhausted after a several-day trip. He had picked up five tuberculosis patients who had been cured with his newest medicine—a thiosemicarbazone that Bayer had named Conteben—and their physician and trucked them to a sanitarium in Flensburg, north, near the Danish border, in an attempt to convince the head of the sanitarium there that this new drug could actually cure TB. Gasoline was in such short supply in postwar Germany that the truck they traveled in was powered by sawdust, wood chips, and coal burned in a firebox. This was the depth to which German technology had been reduced. At least it had the pleasant side effect of keeping them warm on the way: Domagk used the fire to heat wine to ward off the cold; he spent much of the trip passing mulled drinks to the back of the truck. This was not the worst form of transport. On at least one trip after the war, Domagk had traveled by horse and buggy. Much of the nation was still in ruins. Basic services—water, heat, sewage, electricity—were still erratic. Food was scarce. Millions of people had been displaced, and many were still making their way home from massive refugee camps. Four million displaced persons had emigrated from the east to the west of Germany to escape the Russians. Disease, as Domagk had predicted, was everywhere. The Allied occupation forces did what they could, killing lice with DDT and using specially designed bellows to blow disinfectants, including sulfa powder, under the clothes of dispaced persons. Disease rates rocketed regardless.

Before the war 77 of every 100,000 people in Berlin had died from tuberculosis. By 1947, the rate had tripled. Domagk believed that he and the Bayer chemist Robert Behnisch had come up with an answer in the thiosemicarbazones. He had seen Conteben cure dozens of patients. Papers were published in Germany, although in these days communication was not as efficient as it had been and German science was not as well thought of as it had been before the war, so his publications had not attracted a great deal of interest, especially among foreign physicians. Even where his results were known in Germany, physicians were remarkably slow to adopt Conteben. They worried about side effects, they told Domagk, and there was talk of another new drug, called streptomycin, that worked even better. In any case, no drug in history had ever cured tuberculosis, and they had a vested interest in older means of therapy. The only thing that cured TB, they knew, was plenty of bed rest, sunshine, and a long, slow convalescence. The patients had to regain their strength to throw off the disease. TB treatment was accomplished through an extensive system of clinics and sanatoria dedicated to rest cures. Any new drugs threatened a substantial investment in an established mode of business that had been proved to work in many cases.

Domagk called these medical dogmatists the "TB popes." Even when they were shown direct proof, examining patients whom his medicines had cured of even severe pulmonary and bowel tuberculosis (the disease could attack more than just the lungs), he remembered that the physicians "looked at us as if we were either out of our minds or impostors." But he persevered, testing and retesting his drugs, traveling the country with his papers and patients, and he was finally beginning to get the popes to accept his data.

Returning home after the trek to Flensburg, he found a letter addressed to him from the Swedish consulate. Reading it cheered him immensely. Eight years after receiving the Nobel Prize, he was being invited to Stockholm to take part in the official Nobel ceremonies. With Hitler gone he could finally receive his medal.

The ceremonies were only a few weeks away, and there was much to do. He wired the Swedes to let them know that he would be honored to attend. He considered what he would wear. Formal clothes were a necessity, but Domagk's evening clothes had been ruined when American soldiers billeted in his house wore them while playing football. The only other formal wear he had was the suit in which he had been married nearly a quarter of a century earlier, a set of old-fashioned tails that lacked the requisite white tie and vest. The last time he had tried on his marriage suit, in the late 1930s, he remembered he had barely been able to pull the coat halfway around his chest. Now, after years of hardship, he found that it fit perfectly.

Then he started the paperwork. To get a visa, Domagk needed to fill out three questionnaires in two languages and submit them to the local passport office. When he presented the forms, he was told that the occupying forces would also need certification from a priest or physician, as well as a character reference from the police. At the police station, he was told that it would take three months for their background check because of the time required to verify sources at his place of birth. When he told the official that his place of birth was now part of Poland, the policeman laughed and told him, "Then it will take a good deal longer." Domagk reminded the official that this was a special case involving a Nobel Prize and that some degree of haste was needed. He was allowed to substitute a less comprehensive certificate of good conduct covering the twenty years he had lived in the Wuppertal/Elberfeld area.

It seemed that anything related to getting permission to travel out of the country was unnecessarily difficult. It was almost as if the occupying forces were attempting to bury Germany under reams of paper. When he took his growing stack of forms back to the passport office, he was told that he still lacked a health certificate. "The official was quite indignant when I suggested that since I was a doctor maybe I could write one myself," Domagk wrote in his journal. No, the passport official told him, he needed to get an official bill of clean health from a physician in the local community health service. When Domagk dutifully appeared for his examination, he was told that the usual physician was on vacation. He dealt with an assistant, underwent examination, and brought back to the passport office proof that he was free from infectious diseases and parasites. A few days later, he received his visa for Stockholm.

Then the Nobel Committee extended its invitation to include his wife, Gertrude. The Domagks went through the entire process again. Time was growing short when they both secured the necessary permissions and papers. Then Domagk checked with the Swedish consul, who reminded him—"very sympathetically," Domagk remembered—that their train would also be traveling through Denmark. He told them to make certain that their visas were good also for that country. Of course they were not. By now, however, Domagk had learned the system and managed to get the proper paperwork rushed from Hamburg. It arrived the evening before their departure.

A local man had been engaged to pick them up at ten in the morning and drive them to the train station in Düsseldorf, some thirty-five kilometers away. When he did not appear on time, they phoned him. His line was busy. One of Domagk's sons ran to the man's house. The abashed driver sped back and ran to the Domagks' door, apologizing profusely, a coat thrown over his nightshirt, explaining that he had misunderstood the departure time. They could still make it, he vowed, though just barely. He drove them at a hundred kilometers per hour over highways, back roads, and, Domagk remembered, across an open field. They arrived at the station just ten minutes before the international train was scheduled to leave. The Domagks grabbed their bags, hurried to the platform, showed their passports—and were told that this particular train was for international passengers only, no German nationals allowed. They would need to get special permission from the British occupation authorities in order to board. Domagk explained his circumstance, impressed them with the fact that they were dealing with a Nobel Prize winner, and asked them to hold the train until he got back. When he rushed to what he was told was the appropriate office in the station, the British official told Domagk that it would take two weeks to process the necessary paperwork. There was, however, another office in the station where he might get special dispensation for travel emergencies. There, the official pointed, where the line was queued up. Domagk excused his way to the front and finally found someone willing to make something happen quickly. He and his wife were given limited permission to travel as far as

Osnabrück, a German town halfway to the Danish border, to await further permissions.

It was a start. Domagk ran back to the platform with the necessary papers, grabbed his bag, boarded with Gertrude, and began thinking about what to do next. When the train stopped at Osnabrück, they simply stayed in their seats, hoping no one would ask them questions. No one did. By midnight they were in the German town of Flensburg, quite close to Denmark, the town Domagk had visited just a few days before with his truckload of TB patients. Without further permission from the occupying forces, they could not cross the border. In any case there was now another hurdle. The Nobel organization, which was paying for their travel, had arranged for them to pick up tickets for the remainder of the trip at their first stop in Denmark. Domagk, who by now was double-checking everything, persuaded the late-night clerk at the Flensburg station to call up the line to Denmark to make sure that the tickets had arrived. They had not. "So there we were at midnight, my wife and I, together with one suitcase, standing on a station platform," Domagk remembered. "It was raining. The counters were all closed, so we couldn't even deposit our luggage." They were still in Germany, however. People spoke their language, and at least Domagk knew someone in the town; a brother-in-law, Hermann, lived some blocks away from the station. Gerhard and Gertrude tramped there through the dark streets.

Hermann was hospitable and sympathetic. When Domagk told him about his out-of-date evening clothes and the missing white vest, Hermann said not to worry, that he had saved his own evening clothes—a set with a white vest and tie—by hiding them in his car during the occupation. He would be proud if Domagk would wear them at the ceremonies. "At least now, with my borrowed white waistcoat, no matter that my old suit did not look altogether smart, it showed that the spirit was willing," Domagk said.

The next morning they had a good breakfast—there was more food the closer they came to the Danish border—and found to their delight both that their tickets had arrived at the station in Denmark and that their travel had been cleared to Stockholm. The Domagks got on the train, crossed the border, enjoyed their ride through the snowy Danish countryside, and arrived in Copenhagen just after dark. There, when changing trains, Domagk experienced his first taste of what it meant to be a Nobel laureate. He was surrounded by reporters, notebooks out, lightbulbs flashing, asking more questions than Domagk knew how to answer. It was both flattering and disconcerting, and the Domagks beat a hasty retreat to their train compartment. In Stockholm they were treated like visiting royalty, welcomed at the station by Professor Henschen of the Karolinska Institute and a group of young diplomats. One of them noted Domagk's wedding tails, politely saying that the king of Sweden loved old suits like that, then offering to have Domagk's retailored in time for the ceremonies. Domagk refused. "I did not feel a particular need to be seen, as a present-day German, in a new evening suit," Domagk said. "Why conceal reality behind a pretense?"

The next afternoon the Nobel ceremonies, among the most impressive in the modern world, took place in Stockholm's Festival Hall. Gertrude was shown to a special box seat. Gerhard sat with the other Nobelists until a fanfare sounded and they all marched to the platform. It was overwhelming to look out at the two thousand people in the hall, the leading citizens of Sweden, foreign dignitaries, the king and the royal house, the crown prince and princess, all in tails and formal gowns. At the appropriate moment, after hearing his work lauded by a Swedish physician, Domagk rose, walked across the stage, and shook the hand of Sweden's King Gustav V. Domagk, gaunt and fragile-looking, "the relic from 1939," as he described himself, was presented with a gold medal and a hand-lettered diploma. He heard a warm round of applause. He was the first German scientist to appear at the Nobel ceremonies since the war; he was appreciated as a humanitarian and noted for the way he had tried to accept the prize back in 1939 and been punished by the Nazis as a result. The two other German Nobelists from that year, Kuhn and Butenandt—men who had quickly signed their letters of refusal and were rewarded during the war with prestigious research positions in Nazi Germany—were mailed their medals. They never received an invitation to Stockholm. Domagk was a physician, an idealist, a pacifist; he symbolized for the Swedes a healing of wartime wounds. He was not a criminal. He was happy to note the acceptance and warmth he felt everywhere in Sweden.

The rest of the evening was a blur that included seats of honor at the lavish Nobel banquet in the Stockholm City Hall, where the Domagks were joined by members of the royal family, invited to eat food they had not seen since before the war, entertained by leading musicians, and served by university students. "We, the Germans, had to stay away from all the delicacies that were offered because our bodies were not used to that kind of food anymore after all the years of famine," Domagk wrote. "A lot of caricatures were made of us being all meager, sitting side by side with the Americans and the English." The dinner was followed by a serenade from torch-bearing students, then a formal ball. Domagk danced with the beautiful daughter of Sir Edward Appleton, a British winner for physics that year. He and Gertrude stayed until 2:00 A.M., when they were invited to breakfast by the descendants of the Nobel family. "It was very exquisite," he remembered, "but it ended very late."

A few days afterward, Domagk delivered a lengthy Nobel lecture, describing the discovery of Prontosil, now fifteen years old, and his views of what the future held for fighting disease. In it he quickly established both his place in cooperative research—and his sole right to the prize. Domagk was generous to his German coworkers, extending the line of credit all the way through Hörlein to Carl Duisberg, who ran Bayer during the time of the discovery, and Friedrich Bayer, who started the company itself. "The problem of chemotherapy of bacterial infections could be solved neither by the experimental medical research worker nor by the chemist alone, but only by the two together working in very close cooperation over many years," he said. "In particular I must mention the two chemists Dr. Mietzsch and Dr. Klarer, who, thanks to the substances produced by them, enabled me to discover the curative action against bacterial infections after I had worked out and extended step by step, entirely on my own initiative, all possible methods of testing. Convinced that a way could be found, I had persevered with this work over a period of many years, despite all the skepticism prevailing in this field." He then thanked the management of the Bayer Wuppertal-Elberfeld dye factories. His generosity did not extend to the French. Domagk emphasized instead in his speech that the Bayer team was already working on colorless substances when the French announced their findings.

Before launching into a lengthy reprise of his current work—with a special emphasis on tuberculosis—Domagk brought up a troubling problem related to sulfa's use against venereal disease. "For a time people used to speak of a 'lightning cure' for gonorrhea, but ultimately the results became less and less satisfactory although the doses were continually increased," he said. The root of the problem had been seen as early as 1938, when researchers first observed that some strains of gonorrhea were naturally resistant to sulfa drugs. If sulfa drugs were used correctly, the infections were corralled to the point where the body's immune system could clean up whatever germs remained, resistant or not, erasing the infection completely. The problem was that many patients stopped taking the drug as soon as their symptoms disappeared, rather than sticking out the full treatment. That left too many of the most resistant gonorrhea strains alive and ready to spread to others. "The spread of these resistant strains would have been avoided if the few patients who were carriers of them . . . had received careful treatment and had been subjected to clinical supervision until their cure was absolutely certain," Domagk told the Nobel audience in 1947. "But this was not done."

He then warned that the same thing could happen with a new drug, one that had appeared just a few years before but was already beginning to usurp sulfa's place, a medicine called penicillin.

DOMAGK RETURNED TO Germany to more celebrations. The most meaningful was held at the municipal center at Wuppertal, in the building, ironically, where Domagk had been questioned by the Gestapo and held prisoner for a week. Here Philipp Klee, the physician who had first tested Domagk's medicines, paid tribute to his friend and colleague. "There is no better experience for a physician than to see a seemingly incurable illness being cured," he said. "This is the miracle that one always hopes for deep down at the bottom of one's soul." That experience had been made possible for him thanks to Domagk's tireless research, the "manly seriousness with which you looked at your duty and the childlike belief in the infallibility of a successful outcome," Klee said.

Bayer, too, feted the Nobelist, decorating his office at Elberfeld with white lilacs, lilies of the valley, and carnations. Another celebration was organized at the University of Münster, with a musical program and speeches by the local mayor and the head of the university. At the parties at Bayer and Münster both, a special guest attended: Josef Klarer.

At the end of the speech he gave at Münster, Domagk spoke directly about the criminal experiments carried on by German physicians during the war. "Should we not by now take seriously the lessons that were learned from two devastating world wars and the Nuremberg Trials?" he asked. "May the universities in Germany show the right way again in the future, not only to solid knowledge but also to a new humanitarian way, . . . the way to human dignity and the dignity of science."

This was his moment of glory. But by the time Domagk finally received worldwide recognition for his work with sulfa, both he and the drugs he had helped to discover were well on the way to being forgotten.

EPILOGUE

BY 1947, the year Gerhard Domagk traveled to Stockholm to receive his Nobel Prize, sulfa was old news. Public attention had turned quickly to the hot new cure-all for bacterial infections, penicillin. Sulfa's eclipse had started, it could be said, even before it was discovered. In 1928, four years before Klarer made his first test batch of Prontosil, an Australian researcher named Alexander Fleming—another of the cocksure young men in Sir Almroth Wright's House of Lords, the fellow who amused himself by creating petri-dish pictures of ballerinas—noticed that one of his bacterial plates had been spoiled, contaminated with some sort of mold. Then he noted that the mold had the odd effect of clearing a zone around itself in which bacteria did not grow. Fleming figured quite logically that the mold was releasing some sort of substance that hindered the growth of bacteria. He was able to conduct a few tests on his "mould broth," as he called it, but it proved so difficult to purify in amounts required for human medical tests that he dropped the research, turning his attention instead to—what else?—sulfa.

In the late 1930s a pair of British researchers, Howard Florey and Ernst Chain, picked up Fleming's mold broth and made something of it. Florey was encouraged by sulfa's success, which demonstrated for him, as it did for so many others in the 1930s, that magic bullets were indeed possible. He was also inspired to begin working seriously in the field because sulfa fell short of being the perfect magic bullet Ehrlich had sought. Sulfa drugs, mild as they generally were, did have some undesirable side effects and worked well on only some bacteria, leaving many other targets for other medicines. Florey and Chain figured out how to purify the active substance produced by Fleming's mold, how to keep it stable, and how to produce it in sizable amounts. They called the resulting medicine penicillin, after the genus of the mold that produced it, *Penicillium notatum* (from the same Latin root as "pen"; the mold's spores under the microscope look very much like little paintbrushes). Florey and Chain's studies bore fruit during the height of World War II when 100,000 units of penicillin were painstakingly prepared in time to support the D-Day invasion. One year later, penicillin production had risen to 6 billion units, enough to treat a quarter of a million patients. Once penicillin became widely available after the war, everybody started using it. It cured everything that sulfa did (with the exception of dysentery), plus a number of other diseases—including anthrax and syphilis—that sulfa could not touch. Penicillin cleared bacterial infections faster and with generally fewer and milder side effects. By the time Domagk accepted his Nobel Prize, the drug he won it for was becoming a thing of the past.

German researchers were generally less enthusiastic than others around the world about penicillin and the other "antibiotics," as they came to be called, that followed it (the term Domagk preferred for his sulfa medicines, "chemotherapy," is now applied almost exclusively to chemicals used to fight cancer). The Germans, too, pursued penicillin during the war. As early as October 1942 Heinrich Hörlein attended a conference in Potsdam on the antibacterial effects of molds. Another corporate member of IG Farben, Hoechst, started looking for ways to produce penicillin in quantity. One of Hitler's personal physicians oversaw attempts to ramp up production in factories taken from Jews in Germany and Czechoslovakia. Before it could yield appreciable amounts of the medicine, however, the German research program was bombed out of existence. Penicillin was preserved as an asset for the Allies.

In the first two years after the war, many physicians and researchers—including Domagk—continued to advocate the use of sulfa over penicillin.

Domagk's own side-by-side studies showed to his satisfaction that in most infections where both compounds worked, sulfa was just as good as or better than penicillin. He, like many German scientists and others around the world, favored totally synthetic chemicals like sulfa, easily manipulated and made through established chemical processes, over the new mold broths. The biologist Peter Medawar remembered chemists of the day deriding penicillin and its brethren as "obscure, medieval-sounding nostrums" cooked up in cauldrons, "extracts of fungi" more witch's concoction than proper chemistry.

All complaints and most criticisms were drowned, however, in a wave of antibiotics that followed the success of penicillin: streptomycin, chloramphenicol, neomycin, tetracycline, erythromycin, vancomycin, and a host of others. Each helped to increase the total number of diseases that could be cured. Each added to the growing arsenal of weapons available to physicians.

The antibiotic era probably would have happened without sulfa, but it likely would not have happened as quickly. Just as important as its role in curing any disease, sulfa cured the medical nihilism of the 1920s, dissipating the prevailing attitude that chemicals would never be able to cure most diseases. Sulfa proved that magic bullets were possible, encouraged their discovery, established the research methods needed to find them, framed the legal structure under which they would be sold, and created the business model for their development. As one medical historian wrote, "It was the effectiveness of Prontosil, and later other sulfonamide derivatives, that scientists regard as the critical event leading to the resurgence of interest in antibiotics."

The shift from sulfa to the newer antibiotics did not happen overnight. The years immediately after the war saw new sulfa applications—sulfa-impregnated bandages, sulfa-laced nose drops—as well as new molecular forms. Sulphafurazole, for instance, a variation that was far more soluble in urine, became a much-prescribed drug for urinary tract infections. Physicians continued to prescribe sulfa even as the newer antibiotics proved themselves. But the steam finally went out of sulfa in the 1950s as drugmakers began refocusing their efforts on newer, more powerful antibiotics. The explosion of wartime sulfa production—the United States made more than 4,500 tons of sulfa in hundreds of forms in 1943 alone, enough to treat more than 100 million patients—created large stocks, much of which ended up in farmyards, used by veterinarians to cure sick animals, stopping diseases in crowded industrial meat-growing facilities just as it had done in crowded army barracks. The belief spread that sulfa drugs not only kept animals healthy but spurred their growth as well. Sulfa was added to the feed of chickens, pigs, cattle, and fish; used to treat pets; even sprayed on fruits and vegetables to rid them of bacteria before they went to market. For good or ill, its use made possible the close-packed "factory farm" system that produces much of our poultry today. It is still a staple of the farmyard, although recent efforts to lower the amount of antibiotics in meat have tempered its use.

Neither did sulfa entirely disappear from medical use. When Domagk died in the spring of 1964 (ironically, from an infection), the original Prontosil was still being produced and sold, but sulfa's medical use overall had dropped dramatically. In 1957 and 1958, when a brief flurry of interest revolved around new forms of longer-acting and better-tolerated sulfa, a few physicians and drugmakers talked about a "sulfa comeback," but even improved versions could not compete with newer antibiotics. In the late 1960s, there was another burst of interest when researchers found that sulfa used in combination with another drug, trimethoprim, sold under the trade name Bactrim among others, worked extremely well against certain diseases, and is still prescribed for everything from urinary tract infections to the pneumocystis pneumonia that can afflict AIDS patients. Sulfa in one form or another today is used for treating middle ear infections in children, for acne, and for a host of other conditions. But it is now no more than a middling antibiotic among 150 or so on the market.

Like all great discoveries, sulfa engendered a host of unexpected benefits. During the postwar period, Prontosil and its chemical offspring gave birth not just to antibiotics but to other new approaches to disease. Domagk's work, as noted, led to the semithiocarbazones for treating tuberculosis. That was just the beginning. When one doctor observed that patients taking sulfa urinated more often than others, subsequent research led to trying sulfa variants as a diuretic, a medicine used to increase urination and thus to alter the fluid balance in the body. Eventually it led to the thiazide drugs, an important early family of diuretics used to treat hypertension. Understanding sulfa's mode of action—its ability to act as an "antimetabolite" that substituted for a needed foodstuff, starving the target microorganism to death—led to research into other antimetabolites; the most important result was a family of new anticancer drugs. Another line of inquiry that started with sulfa led to antileprosy medications, another to a treatment for diabetes, another to a new line of antimalarials. In all these cases, the starting point was sulfa, but the end point was new kinds of medicine.

Sulfa also changed the *way* drug research was done. Before sulfa, small laboratories followed investigators' hunches and patent-medicine makers cobbled together remedies without testing the results. After sulfa, industrial-scale chemical investigation guided by specific therapeutic goals—the system for finding new medicines pioneered by Hörlein and his Bayer team—became the standard. Successful drugmakers were those who followed the Bayer model, investing enormous sums in basic research at the molecular level, making a vastly increased investment in biochemistry, large-scale animal testing, and toxicology. "No one, so far as I know, had prophesied that substances would act so differently from the antiseptics or in so comparatively gentle a manner," wrote Ronnie Hare, Colebrook's colleague. "But the new ideas had an immense impact on bacteriological thinking. For they not only showed how wrong our previous ideas had been, but they opened up apparently unending avenues of research. . . . It is doubtful whether any of this would have come about had it not been for the discovery of Prontosil."

The new, ever-more-powerful drugs that began flowing from drug companies around the world in the 1950s were clinically tested and legally overseen in ways directly due to the experience with sulfa. The large-scale human tests done with sulfa, from Colebrook's childbed fever patients to tens of thousands of recruits in U.S. Army training bases, helped lay the groundwork for what would from that point on be considered solid evidence of a new drug's effectiveness. The U.S. reform of drug laws following the elixir tragedy institutionalized the expectations for safety and created a new drug-approval process. The legislation, adopted in various forms around the world, also broke the power of the patent-medicine makers and rewarded the new "scientific" breed of drug manufacturer. Today's pharmaceutical industry, with its huge research operations and enormous profits, is the result. The drugmaking industry between 1938, when the new drug laws were enacted in the United States, and 1951, when the era of antibiotics was firmly under way, quickly changed from "a handful of chemical companies with no interest in research and no medical staffs," as historian Richard Hilts put it "to a huge machine that discovered, developed, and marketed drugs of real use in treating disease."

Sulfa, in short, kicked off a revolution in medicine. In its various forms, it had treated untold millions of patients and had saved hundreds of thousands, likely millions, of lives. By 1950, sulfa had been the subject of sixty-five hundred patent applications and the topic of more than fifty-four hundred professional papers in scientific and medical journals. By 1956, just twenty years after Prontosil first became widely available, 90 percent of physicians' prescriptions were written for drugs that did not exist commercially before 1938. Over the same time period, the death rate from childhood diseases dropped more than 90 percent and average life expectancy in the United States increased by more than ten years. Demographers call this "the great mortality transition." It was—as many of those who experienced the effects of its discovery firsthand have commented—"the miracle of miracles in modern medicine." It started with sulfa.

But the effects were not all positive.

In addition to sulfa's great benefits, it also led to the elixir poisonings, the horrible experiments at Ravensbrück, ruined reputations, lawsuits, imprisonment, and suicide. Deeper, longer-lasting, and arguably more significant negative effects are still being felt today. As early as 1939, Perrin Long at Hopkins wrote an article for nurses warning them to keep an eye out for toxic reactions to sulfa, including mania and marked depression—one physician compared the effect of large doses of sulfa on some patients to getting drunk—vomiting, blood-chemistry changes, rashes, and what Long described as "a curious drug fever." None of these were seen as life-threatening, and the answer was simply to lessen or discontinue use. As the war went on, however, physicians began to understand that packing large quantities of sulfa into wounds, as they had done at Pearl Harbor, could increase the risk of long-term liver damage. Very high doses could lead in some cases to anemia or serious kidney conditions. A few physicians came to believe that sulfa slowed tissue healing. It soon became clear that a very small proportion of patients suffered potentially deadly allergic reactions to the drug. At the height of sulfa use, in 1941, twenty-eight people were reported to have died after taking the drug in New York City alone. Given the numbers of people taking sulfa, the rate of severe toxic reactions was still remarkably low. There was, however, clearly a risk, and one that grew with increasing use.

Far more troubling was a second trend. In 1942, military hospitals began seeing a rapid increase in the number of cases in which sulfa no longer seemed effective. An alarming number of soldiers were showing up with formerly sulfa-treatable conditions that now did not respond to the drug. In the late 1930s, for instance, more than 90 percent of soldiers with gonorrhea had responded to sulfa therapy; by 1942, the number was down to 75 percent and falling. In Italy the gonorrhea response rate among British troops fell below 25 percent over the course of a few months. The same was true for other infections. By 1943, the problem of bacterial resistance to sulfa treatment had grown so serious that the U.S. surgeon general called a conference on the topic. Sulfa-resistant strep spread so rapidly that by the war's end it became apparent that sulfa could no longer effectively prevent streptococcal disease among army personnel. One huge set of U.S. Navy tests using sulfa in an attempt to prevent streptococcal infections was stopped in 1945 because of the spread of resistant strains. The Germans, too, stopped using sulfa to prevent gonorrhea in troops because of the increase in resistant strains. The problem was worsened by patients' self-medicating, often in too-small doses for too short a time.

Luckily, by the war's end, penicillin was available to help treat sulfa-resistant infections. Researchers quickly discovered how the bacteria became resistant—there were several biochemical mechanisms employed by different strains. But regardless of how it was done, sulfa researchers realized, misuse of the medicine worsened the problem, most likely by clearing out the bacteria that were sensitive to it, leaving resistant strains to live and take their place. It is likely that the low doses of sulfa given over a period of weeks or months during mass army tests against meningitis, gonorrhea, and strep contributed to the problem of resistance. Despite the problem, sulfadiazine was used for twenty years after the war to treat meningitis in army camps—until, in the early 1960s, sulfa-resistant strains started spreading so rapidly that the fear of a massive outbreak of sulfa-resistant meningitis caused at least one training center to temporarily close. By 1974, a study found that nearly a quarter of all the meningococcus sampled in army trainees was resistant to sulfa.

Sulfa was quickly embraced by the public, used in massive amounts, and often obtained and administered in the early years without any medical supervision. The result was an orgy of misuse, people buying it for the wrong diseases, or using it just long enough to clear up symptoms. Unfortunately, the lessons learned with sulfa did not stop the same thing happening with almost every other antibiotic that followed. The past fifty years have seen a dramatic increase in antibiotic-resistant bacteria of all sorts. Bacteria, it has been discovered, can be promiscuous in the ways they share their DNA with other bacteria and viruses. A gene conferring resistance to one strain of germ can be quickly passed on to other strains, the resistant bacteria surviving the antibiotics used to get rid of them, thriving, and passing the genes on to others. Sometimes the resistant genes for different antibiotics, including sulfa, get packaged together and passed along as a group. Luckily, there are proven ways to stop the spread of resistant bacteria, and physicians are beginning to employ them widely. The most important is simply using antibiotics more carefully, making certain that patients strictly adhere to the dosing schedule, ensuring that the body has a chance to kill off the last resistant bacteria. Regardless, the number of antibiotics on the market, the extent of their use (more than 50 million pounds of antibiotics are produced each year in the United States alone), and their ubiquitous presence (a glass of milk, according to one recent study, can contain minute amounts of up to eighty different antibiotics) have kept the problem of resistance very much alive. The World Health Organization lists antibiotic resistance as one of the three most important public health threats of the twenty-first century.

In the first blush of the sulfa boom in the late 1930s, and again in the first flush of confidence following the fuller introduction of new antibiotics in the 1950s, people got carried away with wonder drugs. They used these unprecedentedly powerful medicines too freely, too carelessly, and too often. Many physicians and patients still do. Sulfa and its children are remarkably seductive drugs.

This new power in medicine led to another effect that seems, at least to many patients, unfortunate. Physicians have changed. Before sulfa, physicians learned their craft in a miscellany of schools, employed a number of approaches, and used techniques developed by a variety of practitioners. In the early 1930s, a quarter of all healers in the United States were homeopaths, osteopaths, chiropractors, Christian scientists, or other "irregular" medical practitioners. They, and the majority of "mainstream" medical practitioners, did what they could with a grab bag of generally weak medicines often made by patent-drug firms or local pharmacists in corner drugstores. Americans had the opportunity to choose from an array of medical approaches, just as they chose from an array of over-the-counter medicines. In both cases quality could vary tremendously—it was an open market, and let the buyer beware. The one thing all physicians had in common, and had held in common since antiquity, was their powerlessness, their inability to fundamentally change the course of most infectious disease. Before sulfa, physicians were by default observers and diagnosers, more able to predict the course of a disease than to do anything about it. Their incomes were, for the most part, modest. Their ambitions were, for the most part, limited. Their goal was often to comfort as much as it was to cure.

Much has changed. Ever-more-powerful drugs, starting with sulfa, led to ever-more-centralized control and oversight of medical care. Physicians accrued increasing power to decide who got what medicine. With burgeoning numbers of new drugs deemed too strong, too potentially toxic, or too easy to abuse to be sold directly to consumers, physicians took an increasing role as gatekeepers. The prescription pad became a forceful tool for determining which medication would be taken by a patient. The right to self-medication, the battle cry of the patent-medicine makers, became more or less a thing of the past. These more powerful physicians were trained in new ways, with a growing emphasis in medical school curricula on the newest, most "scientific" research in molecular biology, physiochemistry, microbiology, and pharmacology. Those physicians who did not master the new methods were gradually marginalized. Homeopathy nearly disappeared in the United States, although a resurgence of interest since the 1960s has kept the study alive, with naturopathy, as "alternative" forms of health care. Where there were once several competing approaches to medicine, there is now only one that matters to most hospitals, insurers, and the vast majority of the public, one that has been shaped to a great degree by the successful development of potent cures that followed the discovery of sulfa drugs. Aspiring caregivers today are chosen as much (or more) for their scientific abilities, their talent for mastering these manifold technological and pharmaceutical advances, as for their interpersonal skills. A century ago most physicians were careful, conservative observers who provided comfort to patients and their families. Today they act: They prescribe, they treat, they cure. They routinely perform what were once considered miracles. The result, in the view of some, has been a shift in the profession from caregiver to technician.

The powerful new drugs changed *how* care was given as well as *who* gave it. Antibiotics made hospitals safer for patients, and alliances were made between mainstream medical schools and hospitals to bring together the strongest medicines with the most skilled practitioners in the most advanced and hygienic care-giving environments. In the 1930s, historian of medicine Paul Starr points out, most medical care took place in patients' homes. Only one physician in sixteen worked at a hospital full-time. Half of all births were home births. The average private physician in 1930 saw about fifty patients each week. By 1950, armed with faster, stronger tools, the average physician was seeing twice that many per week, a rate that has continued to rise. House calls today are almost extinct. More than 90 percent of births take place in hospitals, and it is in hospitals and hospital-associated offices that most physicians do much more of their work. In general, compared to practitioners before the 1930s, physicians today are better trained, better equipped, better able to control the types of medicines that their patients take, much more effective in saving lives, far more harried—and far wealthier.

Finally, the advent of sulfa and the antibiotics that followed changed the way nations approach the control of disease. The period between 1890 and 1930—just after Pasteur and Koch and just before sulfa—is sometimes called "the Golden Era" of public health. This was a time when health professionals understood that germs caused disease but were unable to do anything once patients were infected. The only answer was to prevent the infection. Very effective programs were developed and employed to improve the quality of water, food, and sewage systems; to enhance basic hygiene; and to vaccinate against disease. These programs did wonders to lower disease rates before sulfa. Since 1930, limited health-care funds have increasingly moved away from public health measures toward new drugs and new medical technologies.

Whether these changes prove, in the long term, to be for good or ill remains to be seen. For now, the drugs and techniques are working. We live in a blessed, perhaps all-too-temporary, era in which the invisible predators of the past have been beaten away from our campfires. With yesterday's common killers now relegated to little more than bogeymen in the stories of grandparents and great-grandparents, we suffer instead from diseases of the long-lived conditions that were once considered perquisites of the rich: cancer, obesity, rheumatism, heart disease. Our numbers have risen enormously. The problem now is not staying alive but keeping the earth alive as we overrun it.

If sulfa, the first miracle medicine, shows anything, it is that there is really no such thing as a "miracle" in science. Every great drug discovery (and every modern technological advance) carries with it, like the blood of the Gorgon mentioned in the epigraph that begins this book, two opposing qualities: one positive, healing, and helpful; one negative, often unintended, sometimes deadly. The ancient Greeks understood that. We must remember it, too.

SOURCE NOTES

Rather than extensive footnoting, I have chosen to provide a brief outline of source materials for interested readers. All references refer to entries in the bibliography.

Some sources were tapped for material fairly consistently throughout the book, and rather than repeat them separately under the individual chapter headings below, I have chosen to highlight them once, here, at the beginning.

Wherever possible, I default to primary sources, the raw material of history, original letters, lab notes, diaries, and memoranda. The three major sources for primary papers related to the discovery of the sulfa drugs are the extensive Bayer Archive at Leverkusen, the archive of the Pasteur Institute in Paris, and the Wellcome Library for the History and Understanding of Medicine in London. The Bayer holdings constitute one of the world's richest corporate archives. Here are housed Domagk's laboratory notes, Klarer and Mietzsch's monthly reports, a trove of internal company memos and reports, and the single most important source of information about Domagk's private life: the typescript manuscript of *Lebenserrinnerverrungen,* a long, somewhat eccentric memoir he wrote late in life that remains unpublished. Unfortunately this otherwise superb archive does not offer for public viewing Bayer's top-level administrative records, a corporate policy that limits the chance to completely piece together some puzzles, notably why Bayer hesitated so long before publicly announcing what seemed to be, by most measures, the world's greatest miracle drug. The Pasteur Institute archives are the repository for much firsthand information on the Fourneau laboratory, its workers, and their achievements. Especially important is the Daniel Bovet collection, which includes documents he collected while investigating the Prontosil discovery many years after the fact (Bovet [1988], the result of his labor, is a colorful and useful source to which I returned often). The Wellcome Library's holdings include Leonard Colebrook's papers, a small but valuable source of primary documents that includes his original lab notes and correspondence related to the London Prontosil tests on childbed fever.

The only full-length adult biography of Gerhard Domagk is an admiring study by a former colleague, Ekkehard Grundmann, published in Germany in 2001 and translated into English in 2004. A brief young person's biography, Bankston (2003), offers basic information. Ryan (1993) includes valuable details on Domagk's research and his private life, with a focus on his tuberculosis work. Ryan is especially valuable because it includes very readable translations of many passages from the *Lebenserrinnerverrungen.* Following Domagk's death, significant obituaries were written by Colebrook (1964) and Posner (1971).

The best historical analysis of the scientific work of Domagk, Klarer, and the Bayer research operation, as well as the subsequent research on M&B 693, is the work of science historian John Lesch (see all entries under his name, esp. 1993). Robert Behnisch worked with Domagk at Bayer and describes the laboratory, its scientific approach, and its personalities; Behnisch's recollections (1986) were vital in helping me to re-create the atmosphere and research approach within the company. Bayer's corporate history is recounted in a large, wonderfully illustrated book, Verg (1988), published to celebrate the company's 125th anniversary.

In addition to these often-used sources:

Prologue

A number of books describe the attack on Pearl Harbor, including material about the medical response. I relied on Condon-Rall (1998) for many facts and figures, supplemented by Clarke (2001) and contemporary news sources. The story of John Moorhead can be found on the Web site of the Mamiya Medical Heritage Center (http://hml.org/mmhc); his report for the American Medical Association is published as Moorhead (1942).

Chapter One

The sense of what it was like in a World War I battlefield hospital is drawn primarily from Domagk's memoir, with additional details from Church (1918), Hutchinson (1918), Shay (2002), Higonnet (2001), and MacDonald (1980).

Chapter Two

Information on wound occurrence and methods of treatment during World War I can be found in Hutchinson (1918), Church (1918), Mitchell (1931), MacDonald (1980), Gordon (1993), Higonnet (2001), and Shay (2002). Additional facts about gas gangrene are found in Gordon (1993) and Bhushan (2002). More specific background on Sir Almroth Wright, his research before and during World War I, and his laboratory in the Boulogne Hospital can be found in Dunnill (2000), Colebrook (1954), Cope (1966), Noble (1974), and Heidelberger (1977). The development of antiseptics and the nature of surgery prior to sulfa are described variously in the sources above as well as in Taylor (1942), Galdston (1943), Sokoloff (1949), Hare (1970), Koop (1997), and Zimmerman (2003).

Chapter Three

Basic information on van Leeuwenhoek, Koch, and the other early researchers mentioned in this chapter can be found in a variety of scientific biographical encyclopedias (see, for example, Asimov, 1982); van Leeuwenhoek wrote many fascinating letters, collected and published by the Committee of Dutch Scientists (1939). Louis Pasteur has been the subject of numerous biographies (see Dubos, 1976); the importance of his work and Koch's research as steps toward the discovery of sulfa is also recounted in Taylor (1942) and Galdston (1943). The importance of early discoveries in bacteriology to medicine is highlighted in Williams (1982) and Warner (1986). Further information on the changing attitude of physicians toward bacteriology and drug therapy in the decades prior to the 1930s can be found in Hare (1970), Foster (1970), Lesch (1997), Thomas (1983), Balis (2000), Young (1967), Le Fanu (1999), and Zimmerman (2003).

Chapter Four

The history of Lister and Listerism as it relates to the development of chemotherapy is recounted in Taylor (1942), Galdston (1943), Sokoloff (1949), and Gordon (1993); a number of biographies of Lister are available as well.

Chapter Five

Details of Carl Duisberg's career and the development of the Bayer firm are found in Augustine (1994), Mann and Plummer (1991), and Stokes (1988). More general information on synthetic dyes, the German dye firms, and the formation of IG Farben can be found in Aftalion (1991), Travis (1993), Hayes (2000 and 2001), Lesch (2000), and Higby and Stroud (1997). Further information on the Council of the Gods was obtained from the Bayer Archive; a striking painting of the meeting with Duisberg as its primary focus hangs in the Bayer Kasino at Leverkusen.

Chapter Six

A new biography of Paul Ehrlich, the man with blue, yellow, red, and green fingers, is long overdue. The most recent, Bäumler (1984), a translation of a 1979 German work, is now out of print. Aspects of Ehrlich's approach and achievements are recounted in Mann (1999), Aftalion (1991), Taylor (1942), Zimmerman (2003), Albert (1965), Bowden (2003), De Kruif (1926), Dubos (1941), Galdston (1943), Sokoloff (1949), Foster (1970), Marks (1997), Williams (1982), Liebenau (1987), and Higby and Stroud (1997). More information on syphilis and Salvarsan can be found in Gordon (1993), Thomas (1983), Hare (1970), Quétel (1992), and Tomes (1998). Additional information on Almroth Wright's experience with Optochin is provided in Cope (1966), Noble (1974), Dunnill (2000), and Colebrook (1954).

Chapter Seven

I owe much of my interpretation of Hörlein's role at Bayer to Lesch (1993). Additional information on Hörlein and his research ideas can be found in Ratliff (1937), Silverman (1942), Taylor (1942), and in Hörlein's own words (1936 and 1937). Roehl's career and the research setup he built at Bayer is treated in Lesch (1993), Amyes (2001), Albert (1965), Verg

(1988), Le Fanu (1999) and Behnisch (1986). The holdings of the Bayer Archive were used to provide details about both of these men as well.

Chapter Eight

Calvin Coolidge Jr.'s story was extensively covered in contemporary news reports and retold by Galdston (1943), Gilbert (1998), and Ross (1962). The dangers of strep infection prior to sulfa are detailed in Zimmerman (2003), De Kruif (1932), Silverman (1942), Taylor (1942), and Hare (1955 and 1970). Rebecca Lancefield's strep-typing work is described in McCarty (1987).

Chapter Nine

Leonard Colebrook's career is reviewed in greatest depth in the biography by Noble (1974). See also Colebrook (1954 and 1956); substantive obituaries by Oakley (1971), and in *The Lancet* (1967); as well as descriptions of his work in Parker (1994) and Turk (1994). More on the history of childbed fever, maternity care, early epidemics and treatment, Semmelweis, and Holmes can be found in Nuland (2003), Ayliffe (2003), Loudon (1992, 1995, 2000), Leavitt (1986), Wertz (1989), De Kruif (1936), Churchill (1850), Risse (1999), Semmelweis (1983), Tomes (1998), and Gordon (1993).

Chapter Ten

The information I used to describe the discovery of Prontosil is drawn primarily from lab notebooks, internal reports, and memoranda housed in the Bayer Archive. My work differs appreciably from later reports of the discovery of Prontosil, which were often based on Domagk's first publication (1935) that described only a single experiment. More information can be found in the major sources noted in the introduction to this section, especially Lesch (1993) and Behnisch (1986). Other sources, such as Taylor (1942), Brock (1999), Silverman (1942), Amyes (2001), Galdston (1943), and Northey (1948), include useful information but often fall into the trap of repeating the impression created by Domagk's article that the discovery came in a single great moment, from a single discoverer, via a single experiment. All news, magazine, and book accounts of the discovery should be read critically in light of the more complex story told by the primary documents.

Chapter Eleven

Most of the sources listed for Chapter Ten apply here as well. A number of documents recording the early clinical testing of Streptozon/Prontosil are found in the Bayer Archive; in addition to the standard sources on Domagk, see also Northey (1948), Ryan (1993), and Roberts (1989). For the initial coolness to Domagk's first Prontosil paper, see Hare (1970), Taylor (1942), Dowling (1977), and Hilts (2003).

Chapter Twelve

Heinrich Hörlein's address to the Royal Society of Medicine has been published (Hörlein, 1936). Correspondence with Sir Henry Dale, laboratory notes, and other relevant primary documents relating to Colebrook's introduction to Prontosil are found at the Wellcome Library. Noble (1974) reviews Colebrook's childbed fever work in detail. Hare (1955 and 1970) witnessed (and inadvertently became an early subject of) the London tests of Prontosil, and lived to tell about it with humor, color, and a touch of acid. See references to Chapter Nine for more on the history of childbed fever and its treatment; a number of these sources include information on Colebrook and Kenny's work as well. See also Colebrook and Kenny's original two papers (1936), Ratcliff (1937), Colebrook (1956), and Oakley (1971).

Chapter Thirteen

The archives of the Pasteur Institute house numerous primary documents relating to the work of the Fourneau laboratory, upon which most of this chapter is based. In addition I have relied greatly on Bovet (1988) for details about the group and its discoveries. See also Silverman (1942), Aftalion (1991), Lechevalier (1965), and Hare (1970).

Chapter Fourteen

The two-plus-year gap between the discovery of Prontosil and the first publication from Bayer was the subject of much discussion among scientists in the late 1930s and 1940s. Often it appears that views were colored by the observer's attitude toward Nazi Germany—in other words, a number of French, British, and American writers seem almost eager to assume the worst of the Bayer group (see for instance, Taylor [1942], Silverman [1942], Hare [1970], and the correspondence between Fourneau and Hörlein during the

1930s). I constructed my own interpretation based on a fresh reading of primary documents from the Bayer and Pasteur archives. The debate centers on whether the Bayer researchers discovered the power of pure sulfa and kept it from the world while searching for a more advantageous patent position or whether they failed to discover it at all. Bovet (1988) combed through the Bayer records in the 1980s, decades after his work with Fourneau, and concluded that the Germans missed it. The available evidence, in my view, does not allow a firm conclusion. While a quick review of the lab records seems to indicate that the Bayer team did not know about or test pure sulfa until after the French published their results, too many important documents are missing or unavailable—a few of Klarer's monthly reports, some letters between Fourneau and Hörlein, and virtually all upper-level administrative records, including Hörlein's reports and memoranda to his superiors—to be certain. Careful, repeated reading of the available documents yields even less certainty. Some of the relevant molecules that Klarer noted he was developing, for instance, never appear in Domagk's test reports. There are hints that investigations were undertaken for which I could find no documentation; Hörlein, for instance, later referred to Bayer's work on "colorless" versions of Prontosil prior to 1935, a line of research for which I have been unable to find primary records at the Bayer Archive (other than a very small number of mystery substances, notably Kl-821, that raise more questions than they answer). Hörlein might simply have been trying to save face with his after-the-fact claims; on the other hand, a substantial research record might still be buried in private company documents. There was a great deal of money at stake. IG Farben's changing relationship with the Nazi government is examined most fully in Hayes (2000 and 2001); more can be found in Mann and Plummer (1991) and, more sensationally, Borken (1979). Hörlein's Nazi Party sympathies are examined in the Nuremberg Trial transcripts. I am grateful to Ute Deichmann (2002) for adding to my knowledge of Hörlein's anti-Jewish activities. The story of Domagk's treating his daughter, Hildegarde, with a still-experimental drug has been reported with varying degrees of literary license by different writers, the event shifting in detail and even in year. My version is based on Domagk's medical notes (at the Bayer Archive) and the work of Grundmann (2004), who interviewed Hildegarde's brother about the incident. The early sales success of Prontosil is documented in the financial reports of the Bayer Archive. The importance of

Colebrook's Prontosil session at the Second International Congress of Microbiology is remembered in Galdston (1943), Noble (1974), and Hare (1970). For more on Long and Bliss, see also Silverman (1942) and Balis (2000). For early clinical tests in the United States, Amyes (2001), Ryan (1993), and Dowling (1997).

Chapter Fifteen

I relied to a great extent on contemporary news coverage, especially in the *New York Times,* to reconstruct the story of FDR Jr. See also Cook (1999), Roosevelt (1949, 1989), *Time* (1936), *Newsweek* (Dec. 26, 1936), and *Fortune* (1939).

Chapter Sixteen

For early impressions of the sulfa boom in the United States following the FDR Jr. case, see Kaempffert (1937), Ratcliff (1937), Harding (1938), Spink (1940), Taylor (1942), Silverman (1942), Thomas (1983), and Koop (1997); for later analyses Bickel (1972), Dowling (1977), Werth (1994), and Balis (2000). A complete list of medical and scientific journal articles on Prontosil and Prontosil album (aka Prontylin) through mid-1938 can be found in Winthrop-Stearns, Inc. (1938). Details on Bayer's activities in the United States are based mainly on primary documents from the company archive in Leverkusen. The ongoing feud between Fourneau and Hörlein is documented in the archives at Bayer and the Pasteur Institute, as well as being discussed in Bovet (1988). Long and Bliss's success with meningitis was covered in contemporary news stories (*Fortune* 1939, Ratcliff 1937). The Paris Exposition of 1937 is described in a number of promotional pieces produced by the exposition itself, news stories of the day, and by later historians (see, for example, Peer 1998).

Chapter Seventeen

Major sources of information on the elixir tragedy include Wallace (1937), Young (1967), Jackson (1970), Ballentine (1981), Wax (1995), Balis (2000), and a wealth of contemporary news stories. Massengill (1940) wrote a history of his company, excluding the elixir incident.

Chapter Eighteen

Henry Wallace's (1937) report to Congress makes fascinating reading for anyone interested in how news and politics collide to create public policy—a reaction discussed also in Jackson (1970), Temin (1980), Wax (1995), Hilts (2003), and Balis (2000). Robins (2005) details Royal Copeland's career and role in shepherding through the 1938 act.

Chapter Nineteen

Silverman (1942) and Taylor (1942) describe the meningitis cures in the Sudan. The treatment of lobar pneumonia is reviewed in those books as well as in Thomas (1983) and Dowling (1977); Lesch (1997) is the central source concerning the discovery and impact of M&B 693. For information on the development of second-generation sulfa drugs, I looked to the sources above as well as Sokoloff (1942), Marks (1997), Williams (1982), and the papers of Daniel Bovet at the Pasteur Institute. Woods (1940) is the first paper on sulfa's mode of action; his work and that of Fields is also discussed in Lockwood (1941) Albert (1965), Dowling (1977), Northey (1948), and Le Fanu (1999).

Chapter Twenty

Domagk's memories of the war and his Nobel Prize are found in his unpublished memoir. Many of the translations from the German that I use here come from Ryan (1993); Grundmann (2004) adds context and detail to the events. Hitler's policies regarding the Nobel Prize are detailed in Hargittai (2002) and Crawford (2000). Williams (1982) discusses the sulfa discovery as an example of "industrialized invention." Amyes (2001) includes information on Domagk's other activities during World War II. The Heydrich assassination is described in greatest detail in MacDonald (1998). First-person testimony about the Ravensbrück camp and its experiments with sulfa are found in the transcripts of the Nuremberg "Doctors' Trial" (much of this is online at www.ushmm.org/research/doctors), and in de Gaulle (1998). For more on the camp, see also Annas and Grodin (1992) and Morrison (2000).

Chapter Twenty-One

Domagk's feelings about the Nazi government are known mainly in retrospect; that is to say, he wrote about his unhappiness with the Third Reich mainly after the fact. While there seems little doubt that he was anything but an enthusiastic supporter of the Nazis—he never joined the party and appears to be at best a reluctant participant in his nation's wartime activities—there is also no doubt that he, however reluctantly, also occasionally signed his correspondence "Heil Hitler," often wore a Nazi captain's uniform, and worked in a company that used forced labor in its efforts to ensure a Nazi victory. Like many patriotic Germans during the war, and especially after his run-in with the Gestapo over the Nobel Prize, Domagk seems in general to have taken to heart the old saying "The nail that stands up gets hammered down." The Nazis' general attitudes toward science and medicine are described in Proctor (1999) and Szollosi-Janze (2001); specific relations with IG Farben are detailed in Hayes (2000 and 2001). More on sulfa use during World War II is found in Dowling (1977), Condon-Rall (1998), Hartcup (2000), Coates (1958), and on the surgeon general's military-medicine history Web site (history.amedd.army.mil/default_index2.html). Colebrook's World War II activities are noted in Noble (1974); Fourneau's in the papers at the Pasteur Institute; and the story of Joseph Meister is recounted in every biography of Pasteur. Wheeler Lipes's story was told in many newspapers during the war. I rely primarily on Marghella (2004) and Lipes's own words, available online at the Web site of the Naval Historical Center (www.history. navy.mil/faqs/faq87-3a.htm), and Lipes's obituary in the *Washington Post* (April 19, 2005, p. B6). The military's mass tests of sulfa drugs against meningitis and gonorrhea during the war are detailed in Havens (1963), Sokoloff (1949), Dowling (1977), and Coates (1958). Gordon (1993) tells the story of Boswell's bouts of the disease and the treatments common at the time. The story of Churchill's near death in wartime Carthage has been told a number of times, recently by Lesch (1997), Sakula (2000), Lax (2004), and in the work of John H. Mather, M.D., on the Web site of the Churchill Center (www.winstonchurchill.org).

Chapter Twenty-Two

Most of the information about Domagk's life and work during the late days of the war comes from the basic sources listed at the beginning of this section, especially Grundmann (2004), Ryan (1993), and Domagk's unpublished memoir. Fourneau's postwar imprisonment is discussed in Bovet (1988) and in documents at the Pasteur Institute Archives. Details of Hörlein's trial at Nuremberg from the prosecutors' perspective are found in DuBois (1953); transcripts and summaries of the Farben trial at Nuremberg provide additional information.

Chapter Twenty-Three

The story of Domagk's trip to receive the Nobel Prize is related in his unpublished memoir, much of it told in Ryan (1993), with details from Grundmann (2004). Domagk's Nobel lecture is available, as every Nobel lecture is, on the Internet, http://nobelprize.org. Ryan (1993) is also the best source on Domagk's tuberculosis research; scientific details of Domagk's later work are found in Behnisch (1986).

Epilogue

The story of penicillin's discovery and development has been told most recently by Lax (2004). I also used information from Bickel (1972), Medawar (1990), Hartcup (2000), and Temin (1980). A great deal has been written about the growing issue of antibiotic resistance; see for example Levy (2002) and Shnayerson (2002). Sulfa's side effects can be found in any physician's drug-reference text; historical perspective on side effects is provided by Sokoloff (1949), Taylor (1942), and Marshall (1941). The analysis of changes in medical care spurred by the discovery of sulfa drugs is my own, informed by the sources mentioned above.

BIBLIOGRAPHY

Aftalion, Fred. *A History of the International Chemical Industry.* Philadelphia: University of Pennsylvania Press, 1991.

Ajanki, Tord. *Medicinal Reading: Of Genius, Pure Chance and Dedicated Hard Work.* London: Taylor & Francis, 1995.

Albert, Adrien. *Selective Toxicity.* London: Methuen & Co., 1965.

Amyes, Sebastian. *Magic Bullets, Lost Horizons.* London: Taylor & Francis, 2001.

Annas, George J., and Michael Grodin. *The Nazi Doctors and the Nuremberg Code.* New York: Oxford University Press, 1992.

Asimov, Isaac. *Asimov's Biographical Encyclopedia of Science and Technology* (2nd ed.). Garden City: Doubleday and Co., 1982.

Augustine, Dolores L. *Patricians & Parvenus: Wealth and High Society in Wilhelmine Germany.* Oxford: Berg Publishers, Ltd., 1994.

Ayliffe, Graham, and Mary P. English. *Hospital Infection: From Miasmas to MRSA.* Cambridge: Cambridge University Press, 2003.

Balis, Andrea. *Miracle Medicine* (doctoral dissertation). New York: City University of New York, 2000.

Ballentine, Carol. "Taste of Raspberries, Taste of Death: The 1937 Elixir Sulfanilamide Incident," *FDA Consumer* June 1981.

Bankston, John. *Gerhard Domagk and the Discovery of Sulfa.* Bear, Del.: Mitchell Lane Publishers, 2003.

Bäumler, Ernst. *Paul Ehrlich, Scientist for Life.* New York: Holmes & Meier, 1984.

Behnisch, Robert. *Die Geschichte der Sulfonamidforschung.* Mainz: Med. Pharmazeut. Studienges, e.V., 1986.

———. "From Dyes to Drugs," in Parnham, M. J. and J. Bruinvels, eds., *Discoveries in Pharmacology,* vol. 3. Amsterdam: Elsevier, 1986.
Bhushan, Vikas, ed. *Blackwell's Underground Clinical Series Microbiology,* vol. 1, 3E. London: Blackwells, 2002.
Bickel, Lennard. *Rise Up to Life.* New York: Charles Scribner's Sons, 1972.
Borken, Joseph. *The Crime and Punishment of IG Farben.* New York: Pocket Books, 1979.
Bovet, Daniel. *Une Chimie qui Guerit.* Paris: Editions Payot, 1988.
Bowden, Mary Ellen. *Pharmaceutical Achievers: The Human Face of Pharmaceutical Research.* Philadelphia: Chemical Heritage Press, 2003.
Brock, Thomas D., ed., *Milestones in Microbiology 1546 to 1940.* Washington, D.C.: American Society for Microbiology, 1999.
Business Week. "Brake on Sulfas," Jun. 22, 1941, 86–88.
Church, James Robb. *The Doctor's Part: What Happens to the Wounded in War.* New York: Appleton, 1918.
Churchill, Fleetwood. *Essays on the Puerperal Fever and Other Diseases Peculiar to Women.* Philadelphia; Lea and Blanchard, 1850.
Clarke, Thurston. *Pearl Harbor Ghosts.* New York: Ballantine Books, 2001.
Coates, John Boyd, Jr., ed. *Preventive Medicine in World War II vol. IV.* Washington, D.C.: Office of the Surgeon General, 1958.
Colebrook, Leonard. *Almroth Wright: Provocative Doctor and Thinker.* London: Heinemann, 1954.
———. "The Story of Puerperal Fever 1800 to 1950," *British Medical Journal* (Feb. 4, 1956): 247–262.
———. "Gerhard Domagk," *Biog. Mem. Fellows Roy. Soc.* 10 (1964): 39–50.
———, and Méave Kenny. "Treatment of Human Puerperal Infections, and of Experimental Infections in Mice, with Prontosil," *Lancet* (June 6, 1936): 1279–86.
———. "Treatment with Prontosil of Puerperal Infections Due to Haemolytic Streptococci," *Lancet* (Dec. 5, 1936): 1319–1326.
Committee of Dutch Scientists, eds. *The Collected Letters of Antoni van Leeuwenhoek.* Amsterdam: Swets & Zeitlinger, Ltd., 1939.
Condon-Rall, Mary Ellen, and Albert E. Cowdrey. *The Medical Department: Medical Service in the War Against Japan.* Washington, D.C.: Center of Military History, 1998.
Cook, Blanche Wiesen. *Eleanor Roosevelt,* vol. 2. New York: Viking, 1999.
Cooter, Roger, Mark Harrison, and Steve Sturdy. *Medicine and Modern Warfare.* Amsterdam: Rodopi, 1999.

Cope, Zachary. *Almroth Wright: Founder of Modern Vaccine-Therapy.* London: Thomas Nelson, 1966.

Crawford, Elisabeth. "German Scientists and Hitler's Vendetta Against the Nobel Prizes," *Historical Studies in the Physical and Biological Sciences,* vol. 31, no. 1 (2000): 37–53.

Cunningham, Andrew, and Perry Williams. *The Laboratory Revolution in Medicine.* Cambridge: Cambridge University Press, 1992.

Deichmann, Ute. "Chemists and Biologists during the National Socialist Era," *Ang. Chem. Int. Ed.* 41:1310–28 (2002).

de Gaulle Anthonioz, Geneviève. *The Dawn of Hope: A Memoir of Ravensbrück.* New York: Arcade Publishing, 1998.

De Kruif, Paul. *Microbe Hunters.* New York: Harcourt, Brace and Co., 1926.

———. *Men Against Death.* New York: Harcourt, Brace and Co., 1932.

———. "Why Should Mothers Die?" *Ladies' Home Journal* 53 (three-part series: March, April, May 1936).

———. *The Male Hormone.* New York: Harcourt, Brace and Co., 1945.

Dodds, E. C. "A Review of Recent Progress in the Chemotherapy of Septicemia," *The Practitioner* 137: 719–24 (1936).

Domagk, Gerhard. *Lebenserrinnerverrungen* (undated typescript memoir). Bayer Archive.

———. "Ein Beitrag zur Chemotherapie der bakteriellen Infektionen," *Deutsche Medizinische Wochenschrift* 7:250 (Feb. 15, 1935).

Dowling, Harry F. *Fighting Infection: Conquests of the Twentieth Century.* Cambridge, Mass.: Harvard University Press, 1977.

Drexler, Madeline. *Secret Agents: The Menace of Emerging Infections.* Washington, D.C.: Joseph Henry Press, 2002.

DuBois, Josiah E. Jr. *Generals in Grey Suits.* London: The Bodley Head, 1953.

Dubos, Rene. "The Significance of the Structure of the Bacterial Cell in the Problems of Antisepsis and Chemiotherapy," in Marshall, E. K. Jr., John S. Lockwood, and Rene J. Dubos, eds., *Chemotherapy.* Philadelphia: University of Pennsylvania Press, 1941.

———. *Louis Pasteur, Free Lance of Science.* New York: Charles Scribner's Sons, 1976.

Dunnill, Michael. *The Plato of Praed Street.* London: R. Soc. Med. Press, 2000.

Foster, W. D. *A History of Medical Bacteriology and Immunology.* London: Heinemann, 1970.

Fortune. "Cure by Chemicals," Sept. 1939, 42.

Galdston, Iago. *Behind the Sulfa Drugs*. New York: D. Appleton-Century Co., 1943.

Gilbert, Robert E. *The Mortal Presidency: Illness and Anguish in the White House*. New York: Fordham University Press, 1998.

Gordon, Richard. *The Alarming History of Medicine*. New York: St. Martin's Press, 1993.

Grundmann, Ekkehard. *Gerhard Domagk—der erste Sieger über die Infektionskrankheiten*. Munich: Verlag, 2001.

———. *Gerhard Domagk: The First Man to Triumph Over Infectious Diseases*. Munich: Lit Verlag, 2004.

Hall, Stephen S. *A Commotion in the Blood*. New York: Henry Holt, 1997.

Harding, T. Swann. "Chemotherapy and Prontosil," *Scientific American* (Jan. 1938): 28–29.

Hare, Ronald. *Pomp and Pestilence*. New York: The Philosophical Library, 1955.

———. *The Birth of Penicillin*. Oxford: Allen and Unwin, 1970.

Hargittai, Istvan. *The Road to Stockholm: Nobel Prizes, Science, and Scientists*. Oxford: Oxford University Press, 2002.

Harris, Jerome S., and Henry I. Kohn. "Resistance to Sulfanilyl Derivatives In Vitro and In Vivo," *Science* (July 4, 1936): 11.

Hartcup, Guy. *The Effect of Science on the Second World War*. New York: Palgrave, 2000.

Havens Jr., W. Paul, ed. *Internal Medicine in World War II, Vol. II*. Washington, D.C.: Office of the Surgeon General, 1963.

Hayes, Peter. "I.G. Farben Revisited: Industry and Ideology Ten Years Later," in Lesch, John E., ed., *The German Chemical Industry in the Twentieth Century*. Dordrecht: Kluwer Academic Publishers, 2000.

———. *Industry and Ideology*. Cambridge: Cambridge University Press, 2001.

Heidelberger, Michael. "A 'Pure' Organic Chemist's Downward Path." *Perspectives in Biology and Medicine*, 1981.

Higby, Gregory J., and Elaine C. Stroud, eds. *The Inside Story of Medicines: A Symposium*. Madison, Wis.: American Institute of the History of Pharmacy, 1997.

Higonnet, Margaret. *Nurses at the Front: Writing the Wounds of the Great War*. Boston: Northeastern University Press, 2001.

Hilts, Philip J. *Protecting America's Health*. New York: Alfred A. Knopf, 2003.

Hogben, Lancelot. *Science for the Citizen.* London: Allen & Unwin Ltd., 1938.

Hörlein, Heinrich. "The Chemotherapy of Infectious Diseases Caused by Protozoa and Bacteria," *Proc. Royal Soc. Med.* 29:313–24 (1936).

———. "The Development of Chemotherapy for Bacterial Diseases," *The Practitioner* 139:635–49 (1937).

Hutchinson, Woods. *The Doctor in War.* Boston: Houghton Mifflin, 1918.

Jackson, Charles O. *Food and Drug Legislation in the New Deal.* Princeton: Princeton University Press, 1970.

Kaempffert, Waldemar. *New York Times* XII; 6:2 (Oct. 24, 1937).

Karlen, Arno. *Man and Microbes.* New York: Touchstone, 1995.

Koop, C. Everett. "Medicines in American Society—A Personal View," in Higby, Gregory, and Elaine Stroud, eds. *The Inside Story of Medicines: A Symposium.* Madison, WI: Amer. Inst. of the Hist. of Pharmacy, 1997.

Lancet. "Leonard Colebrook," Oct. 7, 1967, 783–84.

Lax, Eric. *The Mold in Dr. Florey's Coat.* New York: Henry Holt, 2004.

Le Fanu, James. *The Rise and Fall of Modern Medicine.* New York: Carroll & Graf, 1999.

Leavitt, Judith Walzer. *Brought to Bed: Childbearing in America 1750–1950.* New York: Oxford University Press, 1986.

Lechevalier, Hubert, and M. Solotorovsky, eds. *Three Centuries of Microbiology.* New York: McGraw-Hill, 1965.

Lesch, John E. "Chemistry and Biomedicine in an Industrial Setting: The Invention of the Sulfa Drugs," in Seymour Mauskopf, ed., *Chemical Sciences in the Modern World.* Philadelphia: University of Pennsylvania Press, 1993.

Lesch, John E. "The discovery of M&B 693 (sulfapyridine)," in Higby, Gregory J., and Elaine C. Stroud, eds., *The Inside Story of Medicines: A Symposium.* Madison, Wisc.: American Institute of the History of Pharmacy, 1997, pp. 101–19.

———., ed. *The German Chemical Industry in the Twentieth Century.* Dordrecht: Kluwer Academic Publishers, 2000.

Levy, Stuart B. *The Antibiotic Paradox* (2nd ed.). New York: Perseus, 2002.

Liebenau, Jonathan. *Medical Science and Medical Industry.* Baltimore: Johns Hopkins University Press, 1987.

Lockwood, John S. "Progress Toward an Understanding of the Mode of Chemotherapeutic Action of Sulfonilamide Compounds," in Marshall, E.

K., Jr., John S. Lockwood, and Rene J. Dubos, eds., *Chemotherapy.* Philadelphia: University of Pennsylvania Press, 1941, pp. 9–28.

Long, Perrin H. "The Progress of Science: Award of the Nobel Prize in Physiology and Medicine to Dr. Gerhard Domagk," *Scientific Monthly* (50): 83–84, 1940.

Loudon, Irvine. *Death in Childbirth.* Oxford: Clarendon Press, 1992.

———. *Childbed Fever: A Documentary History.* New York: Garland, 1995.

———. *The Tragedy of Childbed Fever.* Oxford: Oxford University Press, 2000.

MacDonald, C. A. *The Killing of Reinhard Heydrich.* New York: Da Capo, 1998.

MacDonald, Lyn. *The Roses of No Man's Land.* London: Michael Joseph, 1980.

Maehl, William H. *Germany in Western Civilization.* Tuscaloosa: University of Alabama Press, 1979.

Mann, Charles C., and Mark L. Plummer. *The Aspirin Wars.* New York: Alfred A. Knopf, 1991.

Mann, John. *The Elusive Magic Bullet.* Oxford: Oxford University Press, 1999.

Marghella, Pietro D. "World War II Submarine 'Surgeon.'" *Proc. US Naval Inst.* 130(12), (Dec. 2004).

Marks, Harry M. *The Progress of Experiment.* Cambridge: Cambridge University Press, 1997.

Marshall, E. K., Jr. "The Present Status and Problems of Bacterial Chemotherapy," *Science* 91: 345–50 (1936).

———. "The Pharmacology of Sulfanilamide and Its Derivatives," in Marshall, E. K., Jr., John S. Lockwood, and Rene J. Dubos, eds., *Chemotherapy.* Philadelphia: University of Pennsylvania Press, 1941.

Massengill, Samuel Evans. *A Sketch of Medicine and Pharmacy and a View of Its Progress by the Massengill Family from the Fifteenth to the Twentieth Century.* Bristol, Tenn.: S. E. Massengill Co., 1940.

McCarty, Maclyn. "Rebecca Craighill Lancefield," *Biographical Memoirs* 57 (1987): 227–241.

Medawar, Peter B. *The Threat and the Glory.* New York: HarperCollins, 1990.

Miller, Lois Mattox. "Sulfa-Miracles," *Hygeia,* Sept. 1940.

Mitchell, T. J. *Medical Services: Casualties and Medical Statistics of the Great War.* London: Imperial War Museum, 1931.

Moorhead, John J. "Surgical Experience at Pearl Harbor," *JAMA* 118 (9): 712–14 (Feb. 28, 1942).

Morrison, Jack G. *Ravensbrück: Everyday Life in a Women's Concentration Camp 1939–45*. Princeton: Markus Wiener, 2000.

Nathan, Otto, and Heinz Norden, eds. *Einstein on Peace*. New York: Schocken Books, 1960.

Newsweek. "Streptococci: A Reddish Dye Comes to the Aid of Childbirth," June 27, 1936.

———. "Prontylin: New Drug Arrests Roosevelt Jr.'s Sinus Trouble," Dec. 26, 1936.

Noble, W. C. *Coli: Great Healer of Men*. London: Heinemann, 1974.

Northey, Elmore H. *The Sulfonamides and Allied Compounds*. New York: Reinhold Publishing Corp., 1948.

Nuland, Sherwin B. *Doctors: The Biography of Medicine*. New York: Alfred A. Knopf, 1988.

———. *The Doctors' Plague*. New York: Norton, 2003.

Oakley, C. L. "Leonard Colebrook," *Biog. Mem. Fellows Roy. Soc.* 17 (1971): 90–138.

Parker, M. T. "Leonard Colebrook and his Family," *J. Hosp. Inf.* 28:81–90 (1994).

Peer, Shanny. *France on Display*. Albany: State University of New York Press, 1998.

Piel, Gerard. *Science in the Cause of Man*. New York: Alfred A. Knopf, 1962.

Podolsky, M. Lawrence. *Cures Out of Chaos*. Amsterdam: Overseas Press Association, 1997.

Posner, Erich. "Gerhard Domagk," *The Dictionary of Scientific Biography IV* (pp. 153–56). New York: Charles Scribner's Sons, 1971.

Proctor, Robert N. *The Nazi War on Cancer*. Princeton: Princeton University Press, 1999.

Pully, Pete, and Frank W. DeFriece. *Massengill Brothers Company and the S. E. Massengill Company, 1897–1971*. Knoxville: Tennessee Valley Publishing, 1996.

Quétel, Claude. *The History of Syphilis*. Baltimore: Johns Hopkins University Press, 1992.

Ratcliff, J. D. "Magic Dye," *Collier's*, April 3, 1937, 62.

Reister, Frank A. *Medical Statistics in World War II*. Washington, D.C.: Office of the Surgeon General, 1975.

Renneberg, Monika, and Mark Walker, eds. *Science, Technology and National Socialism*. Cambridge: Cambridge University Press, 1994.

Risse, Günter B. *Mending Bodies, Saving Souls: A History of Hospitals*. Oxford: Oxford University Press, 1999.

Roberts, Royston. *Serendipity: Accidental Discoveries in Science*. New York: Wiley Science Editions, 1989.

Robins, Natalie. *Copeland's Cure*. New York: Knopf, 2005.

Rocco, Fiammetta. *The Miraculous Fever-Tree*. New York: HarperCollins, 2003.

Roosevelt, Eleanor. *This I Remember*. New York: Harper, 1949.

———. *My Day* (edited by Rochelle Chadakoff). New York: Pharos Books, 1989.

Ross, Ishbel. *Grace Coolidge and Her Era*. New York: Dodd, Mead & Co., 1962.

Rothman, David J. *Strangers at the Bedside*. New York: Basic Books, 1991.

Ryan, Frank. *The Forgotten Plague*. Boston: Little, Brown and Co., 1993.

Sakula, Alex. "Churchill in Carthage, 1943: Dr. Evan Bedford's War Diary," *Journal of Medical Biography* 8:241–43 (2000).

Schnitzer, R. J., and Frank Hawking. *Experimental Chemotherapy*, vol. 1. New York: Academic Press, 1963.

———. *Experimental Chemotherapy*, vol. 2. New York: Academic Press, 1964.

Self, Sidney B. "Sulfa Comes Back," *Science Digest* 78–81.

Semmelweis, Ignaz. *The Etiology, Concept, and Prophylaxis of Childbed Fever*. Madison, Wis.: University of Wisconsin Press, 1983.

Shay, Michael E. *A Grateful Heart: The History of a World War I Field Hospital*. Westport, Conn.: Greenwood Press, 2002.

Shnayerson, Michael, and Mark Plotkin. *The Killers Within: The Deadly Rise of Drug-Resistant Bacteria*. Boston: Little, Brown, 2002.

Shryock, Richard Harrison. *The Development of Modern Medicine*. London: Gollancz, 1948.

Silverman, Milton. *Magic in a Bottle*. New York: Macmillan, 1942.

Sneader, Walter. *Drug Discovery: A History*. Chichester: John Wiley & Sons, 2005.

Sokoloff, Boris. *The Miracle Drugs*. New York: Ziff-Davis, 1949.

Spink, Wesley W. "Sulfanilamide, Master Germ Killer," *Science Digest*, Summer 1940, 42–46.

Starr, Douglas. *Blood: An Epic History of Medicine and Commerce*. New York: Quill, 2000.

Starr, Paul. *The Social Transformation of American Medicine*. New York: Basic Books, 1982.

Stevenson, Lloyd G. *Nobel Prize Winners in Medicine and Physiology, 1901–1950*. New York: H. Schuman, 1953.

Stokes, Raymond G. *Divide and Prosper: The Heirs of I. G. Farben Under Allied Authority 1945–1951*. Berkeley: University of California Press, 1988.

Szollosi-Janze, Margit, ed. *Science in the Third Reich*. New York: Berg, 2001.

Taylor, F. Sherwood. *The Conquest of Bacteria*. New York: Philosophical Library and Alliance Book Corp., 1942.

Temin, Peter. *Taking Your Medicine: Drug Regulation in the United States*. Cambridge, Mass.: Harvard University Press, 1980.

Thomas, Lewis. *The Youngest Science*. New York: Viking, 1983.

Thompson, John D., and Grace Goldin. *The Hospital: A Social and Architectural History*. New Haven: Yale University Press, 1975.

Time. "Prontosil," Dec. 28, 1936, 21.

———. "New Wonder Drug," Feb. 23, 1937, 67.

———. "Dangerous Drug," April 6, 1937, 61–62.

———. "Again, Sulfanilamide," Aug. 30, 1937, 61.

Tomes, Nancy. *The Gospel of Germs*. Cambridge, Mass.: Harvard University Press, 1998.

Travis, Anthony S. *The Rainbow Makers*. Bethlehem, Pa.: Lehigh University Press, 1993.

Tréfouël, Jacques. "Ernest Fourneau (1872–1949)," *Bull. Nat. Acad. Med.* 31:589 (1949).

Turk, J. L. "Leonard Colebrook: the chemotherapy and control of streptococcal infections," *J. Roy., Soc. Med.* 87:727–28 (1994).

Verg, Erik, ed. *Milestones: The Bayer Story 1863–1988*. Leverkusen: Bayer AG, 1988.

von Humboldt, Alexander. *Gerhard Domagk 1895–1964, Lebenserinnerungen in Bildern und Texten*. Leverkusen: Bayer AG, 1995.

Wallace, Henry A. "Elixir Sulfanilamide: Letter from the Secretary of Agriculture Transmitting in Response to Senate Resolution No. 194, a Report on Elixir Sulfanilamide-Massengill," *Congressional Record,* Senate Doc. #124, Nov. 16 (calendar day, Nov. 26) 1937. Washington, D.C.: U.S. Govt. Printing Office, 1937.

Warner, John Harley. *The Therapeutic Perspective*. Cambridge, Mass.: Harvard University Press, 1986.

Wax, Paul M. "Elixirs, Diluents, and the Passage of the 1938 Federal Food, Drug and Cosmetic Act," *Annals of Internal Medicine* 122(6): 456–61 (1995).

Weatherall, M. *In Search of a Cure*. Oxford: Oxford University Press, 1990.

Werth, Barry. *The Billion Dollar Molecule*. New York: Simon & Schuster, 1994.

Wertz, Richard W., and Dorothy C. Wertz. *Lying-In: A History of Childbirth in America*. New Haven: Yale University Press, 1989.

Williams, Trevor. *A Short History of Twentieth Century Technology*. Oxford: Clarendon Press, 1982.

Winthrop-Stearns, Inc. *Prontosil & Prontylin Bibliography* New York: Winthrop Chem. Co., 1938.

Woods, D. O. 1940. "The relation of p-aminobenzoic acid to the mechanism of the action of sulphanilamide," *Brit. J. Exp. Path.* 21: 74–90.

Young, James Harvey. *The Medical Messiahs*. Princeton: Princeton University Press, 1967.

———. "Sulfanilamide and Diethylene Glycol," in Parascandola, John, and James C. Whorton, eds., *Chemistry and Modern Society*. Washington, D.C.: American Chemical Society, 1983.

Zimmerman, Barry E. *Killer Germs*. New York: Contemporary Books, 2003.

INDEX